Cryptosporidium: Answers to Questions Commonly Asked by Drinking Water Professionals

The mission of the AWWA Research Foundation is to advance the science of water to improve the quality of life. Funded primarily through annual subscription payments from over 900 utilities, consulting firms, and manufacturers in North America and abroad, AWWARF sponsors research on all aspects of drinking water, including supply and resources, treatment, monitoring and analysis, distribution, management, and health effects.

From its headquarters in Denver, Colorado, the AWWARF staff directs and supports the efforts of over 500 volunteers, who are the heart of the research program. These volunteers, serving on various boards and committees, use their expertise to select and monitor research studies to benefit the entire drinking water community.

Research findings are disseminated through a number of technology transfer activities, including research reports, conferences, videotape summaries, and periodicals.

Cryptosporidium: Answers to Questions Commonly Asked by Drinking Water Professionals

Prepared by:
Michelle M. Frey
Hagler Bailly Services, Inc., 1881 9th Street, Suite 201
Boulder, CO 80302
Carrie Hancock
CH Diagnostics, Inc., 214 S.E. 19th Street
Loveland, CO 80537
Gary S. Logsdon
Black & Veatch, 10250 Alliance Road, Suite 101
Cincinnati, OH 45242

Sponsored by:
AWWA Research Foundation
6666 West Quincy Avenue
Denver, CO 80235-3098

Published by the
AWWA Research Foundation and
American Water Works Association

Disclaimer

This study was funded by the AWWA Research Foundation (AWWARF). AWWARF assumes no responsibility for the content of the research study reported in this publication or for the opinions or statements of fact expressed in the report. The mention of trade names for commercial products does not represent or imply the approval or endorsement of AWWARF. This report is presented solely for informational purposes.

Library of Congress Cataloging-in-Publication Data
Frey, Michelle.
Cryptosporidium : answers to questions commonly asked by drinking water professionals / prepared by Michelle M. Frey, Carrie Hancock, and Gary S. Logsdon.
p. xvi,72 cm. 21.5×28
Includes bibliographical references (p.).
ISBN 0-89867-937-0
1. Cryptosporidium—Control. 2. Drinking water—Contamination. 3. Cryptosporidiosis. I. Hancock, Carrie. II. Logsdon, Gary S. III. Title.
TD427.C78F74 1998
628.1′62—dc21 97-32065
CIP

Printed in the U.S.A.

ISBN 0-89867-937-0

Printed on recycled paper.

CONTENTS

TABLES AND FIGURES

Tables

Figures

FOREWORD

The AWWA Research Foundation is a nonprofit corporation that is dedicated to the implementation of a research effort to help utilities respond to regulatory requirements and traditional high-priority concerns of the industry. The research agenda is developed through a process of consultation with subscribers and drinking water professionals. Under the umbrella of a Strategic Research Plan, the Research Advisory Council prioritizes the suggested projects based upon current and future needs, applicability, and past work; the recommendations are forwarded to the Board of Trustees for final selection. The foundation also sponsors research projects through unsolicited proposal process; the Collaborative Research, Research Applications, and Tailored Collaboration programs; and various joint efforts with organizations such as the U.S. Environmental Protection Agency, the U.S. Bureau of Reclamation, and the Association of California Water Agencies.

This publication is a result of one of these sponsored studies, and it is hoped that its findings will be applied in communities throughout the world. The following report serves not only as a means of communicating the results of the water industry's centralized research program but also as a tool to enlist the further support of the nonmember utilities and individuals.

Projects are managed closely from their inception to the final report by the foundation's staff and large cadre of volunteers who willingly contribute their time and expertise. The foundation serves a planning and management function and awards contracts to other institutions such as water utilities, universities, and engineering firms. The funding for this research effort comes primarily from the Subscription Program, through which water utilities subscribe to the research program and make an annual payment proportionate to the volume of water they deliver and consultants and manufacturers subscribe based on their annual billings. The program offers a cost-effective and fair method for funding research in the public interest.

A broad spectrum of water supply issues is addressed by the foundation's research agenda: resources, treatment and operations, distribution and storage, water quality and analysis, toxicology, economics, and management. The ultimate purpose of the coordinated effort is to assist water suppliers to provide the highest possible quality of water economically and reliably. The true benefits are realized when the results are implemented at the utility level. The foundation's trustees are pleased to offer this publication as a contribution toward that end.

One of the most critical areas of research for drinking water utilities is the control of *Cryptosporidium* in water supplies. Research efforts have expanded in response to this clear industry need, and consequently, a rapid pace is being set for the generation of new information. Drinking water professionals often are overwhelmed by the breadth and depth of available research on *Cryptosporidium* and find themselves without the necessary resources to compile, synthesize, and apply this important information. This report provides these professionals with concise and direct answers to many questions concerning *Cryptosporidium* and drinking water. Further technical reference material can be found in AWWARF's companion report, *Critical Evaluation of* Cryptosporidium *Research and Research Needs* (Frey et al. 1997), which presents a synthesis of *Cryptosporidium* research findings, the status of on-going research efforts, and directions for future research activities.

George W. Johnstone
Chair, Board of Trustees
AWWA Research Foundation

James F. Manwaring, P.E.
Executive Director
AWWA Research Foundation

ACKNOWLEDGMENTS

The authors of this report would like to acknowledge the contributions made by researchers worldwide in support of this effort. Special thanks are due to Jeffrey Oxenford, AWWARF project manager for this study. In particular, we appreciate the thoughtful comments provided by Mark Knudsen, Roger Edwards, and Ross Walker with the Portland Water Bureau, Portland, Ore.; Jennifer Clancy with Clancy Environmental Consultants, St. Albans, Vt.; Gordon Finch with the University of Alberta, Edmunton; Andrew DeGraca with the San Francisco Public Utilities Commission; and Frances Taylor with the San Francisco Department of Health. We also wish to acknowledge the assistance of Anita Heffernan, Black and Veatch, Aurora, Colo., and Shana DeBault, Hagler Bailly Services, Inc., Boulder, Colo., in the preparation of this manuscript.

EXECUTIVE SUMMARY

Cryptosporidium in drinking water is a critical concern to water utilities worldwide as the consequences of this public health issue become increasingly understood. Only a decade ago, *Cryptosporidium* was not on the horizon as a priority for public health protection in the United States. With the increasing incidence of waterborne cryptosporidiosis outbreaks, the public health and drinking water communities initiated numerous research efforts. In fact, the American Water Works Association Research Foundation (AWWARF) expended nearly $9 million (M$) in *Cryptosporidium* research from 1989 to 1996, resulting in a total research value of 18.6 M$ when in-kind contributions were considered. This considerable investment by AWWARF was made in response to an AWWARF survey in which U.S. drinking water utilities ranked *Cryptosporidium* as the highest research priority for the drinking water community.

With the plethora of new information available on *Cryptosporidium*, utilities have found themselves lacking the resources to fully compile, synthesize and apply the pertinent research findings. This project was undertaken to provide such a resource to the drinking water community. A workshop of utility managers was conducted to solicit input on the critical issues and questions facing drinking water utilities concerning *Cryptosporidium*. The managers in the workshop identified key *Cryptosporidium*-related questions, and based on a critical review and synthesis of available literature and on-going research efforts, this report answers those questions.

The questions and answers presented in this report are organized by the following topics:

- Public Health and Awareness
- Occurrence in Source Waters
- Monitoring and Analysis
- Treatment Effectiveness: Removal and Inactivation
- Public Information and Management

In addition, sources for more information on *Cryptosporidium* and its related research are provided, including web page locations, hotline telephone numbers, and a list of suggested readings.

This report has been prepared to address the information needs of a broad range of people, including utility managers and operators, consulting engineers, public health officials, public liaison or involvement personnel, and the general public. Given this targeted audience, the answers are intended to provide sufficient depth to fully address the questions without overwhelming the reader with details. For further technical information on *Cryptosporidium*-related research, a companion report is available from AWWARF and AWWA, *Critical Evaluation of* Cryptosporidium *Research and Research Needs* (Frey et al., 1997).

The science on *Cryptosporidium* is rapidly expanding, and utilities will continue to face difficult questions in the future. A reevaluation of the answers presented in this report will be an important component in keeping apprised of the state-of-the-art knowledge concerning *Cryptosporidium*.

CHAPTER 1
INTRODUCTION

Cryptosporidium in drinking water is a critical concern to water utilities worldwide as the consequences of this public health issue become increasingly understood. Only a decade ago, *Cryptosporidium* was not on the horizon as a priority for public health protection in the United States. With the increasing incidence of waterborne cryptosporidiosis outbreaks, the public health and drinking water communities initiated numerous research efforts. In fact, the American Water Works Association Research Foundation (AWWARF) expended nearly $9 million (M$) in *Cryptosporidium* research from 1989 to 1996, resulting in a total research value of 18.6 M$ when in-kind contributions were considered. This considerable investment by AWWARF was made in response to an AWWARF survey in which U.S. drinking water utilities ranked *Cryptosporidium* as the highest research priority for the drinking water community.

With the plethora of new information available on *Cryptosporidium*, utilities have found themselves lacking the resources to fully compile, synthesize and apply the pertinent research findings. This project was undertaken to provide such a resource to the drinking water community. A workshop of utility managers was conducted to solicit input on the critical issues and questions facing drinking water utilities concerning *Cryptosporidium*. The managers in the workshop identified key *Cryptosporidium*-related questions, and based on a critical review and synthesis of available literature and on-going research efforts, this report answers those questions.

The questions and answers presented in this report are organized by the following topics:

- Public Health and Awareness
- Occurrence in Source Waters
- Monitoring and Analysis
- Treatment Effectiveness: Removal and Inactivation
- Public Information and Management

In addition, sources for more information on *Cryptosporidium* and its related research are provided, including web page locations, hotline telephone numbers, and a list of suggested readings. A complete listing of the AWWARF reports on *Cryptosporidium* is also available (Appendix A).

This report has been prepared to address the information needs of a broad range of people, including utility managers and operators, consulting engineers, public health officials, public liaison or involvement personnel, and the general public. Given this targeted audience, the answers are intended to provide sufficient depth to fully address the questions without overwhelming the reader with details. For further technical information on *Cryptosporidium*-related research, a companion report is available from AWWARF and AWWA , *Critical Evaluation of* Cryptosporidium *Research and Research Needs* (Frey et al., 1997).

The science on *Cryptosporidium* is rapidly expanding, and utilities will continue to face difficult questions in the future. A reevaluation of the answers presented in this report will be an important component in keeping apprised of the state-of-the-art knowledge concerning *Cryptosporidium*.

CHAPTER 2

PUBLIC HEALTH AND AWARENESS

At the turn of the 19th century, the ability to reduce disease epidemics was linked to treating the drinking water supply serving urban communities. Since that time, public water systems have had as their primary mission the provision of a safe and plentiful supply of potable water. Today, this challenge is still foremost in the success of a public water system, and awareness of the public health concerns related to drinking water supplies is fundamental. This section poses and answers several questions related to the public health issue of *Cryptosporidium* exposure via drinking water.

What is *Cryptosporidium,* and where does it originate in the environment?

Cryptosporidium is a single-celled organism that can be seen with the aid of a microscope. It reproduces within the gut of an animal host where it also gleans its nutrition. One step during the life cycle of *Cryptosporidium* involves the maturation into resistant cells called *oocysts*. Oocysts can be shed from the animal host during bowel movements. Once shed from the animal, oocysts may be dispersed by water, by aerosols or via the animal itself or other animals. Oocysts can survive a variety of temperatures, wet and dry environments, and exposure to chemicals. If surviving oocysts are consumed by a suitable animal host, they can infect the animal and will continue their life cycle by reproducing within the animal's gut. Therefore, *Cryptosporidium* is only introduced into the environment within animal feces, including human feces; it does not reproduce outside of an animal.

Can *Cryptosporidium* make people ill, and if so, who is at risk from *Cryptosporidium*?

The disease caused by *Cryptosporidium* is called *cryptosporidiosis* and is caused by infection with or by *Cryptosporidium* oocysts. People can be exposed to oocysts from other people, animals, drinking or recreational water, fresh food, soils, and any surface that has not been sanitized after exposure to feces. For example, numerous outbreaks of cryptosporidiosis have been linked to day care center and swimming pool activities where family members also become ill due to subsequent human-to-human transmission from the exposed children or participants.

Cryptosporidium can make people ill; however, not every person who becomes infected with *Cryptosporidium* feels ill or notices symptoms. Other people, particularly those with compromised or suppressed immune systems, can have severe life-threatening cryptosporidiosis. Symptoms can range from a week of mild diarrhea to incapacitating diarrhea that does not stop, along with cramps, loss of appetite, weight loss, nausea, and low-grade fever. Because a safe and effective treatment for cryptosporidiosis has not been developed, special care should be taken by people suffering from AIDS, receiving certain cancer treatments, having had organ transplants, or living with a lowered disease resistance due to illness, age (typically children less than six years old and the elderly), or undernourishment.

The Centers for Disease Control (CDC 1995) recommend that persons believing they may have cryptosporidiosis should talk about testing and treatment with their health care providers. Prevention of cryptosporidiosis can be achieved only by eliminating all routes of exposure. The CDC (1995) specifically recommend that persons concerned about cryptosporidiosis (1) maintain stringent hygienic habits like thorough hand-washing after contact with other people or animals, (2) consume sanitary food and beverages, (3) not practice oral sex (i.e., not engage in sexual activity that may result in contamination of the hands or mouth with fecal matter), and (4) boil or filter drinking water through an at least *absolute* 1-μm-pore-size filter. It is important to remember to also observe care with “cryptic” ingestion of water that typically occurs via ice, fountain soft drinks, reconstituted juices, toothbrushing, and recreational water ingestion. Whatever manner of decontaminating water is used at home must be applied away from home just as diligently.

Have illnesses resulted from *Cryptosporidium* in drinking water?

Waterborne outbreaks of cryptosporidiosis have been identified within the United States and other countries in communities served by public water supply systems (Table 2.1). Direct evidence of the water supply being the exposure source for these outbreaks has not always been available.

When the outbreaks are evaluated to identify the causes, monitoring of the finished water for *Cryptosporidium* levels either preceding or during the outbreak period would be necessary to provide a direct link. However, this information is often unavailable, or if some is available, it has been limited in quantity and of questionable quality. In some outbreaks, though, analysis of the risk factors associated with infection leads epidemiologists to conclude that drinking water was the probable source of infection.

How are disease outbreaks identified, and are there steps a utility can take to identify potential outbreak events?

Disease "outbreaks" are defined by the Centers for Disease Control (CDC) and most epidemiologists as an occurrence of disease greater than would be expected. The key, then, is for a community to define a "baseline" level of disease occurrence. Both the establishment of a disease occurrence baseline and the identification of outbreak events depend on active reporting of illnesses by physicians and a surveillance or tracking system to monitor for deviations from a baseline level of disease occurrence. In the case of cryptosporidiosis, diagnosed cases should be reported to CDC. However, medical providers often do not test for the cause of diarrhea that goes away by itself, resulting in an underreporting of actual disease events.

Those communities with disease tracking or surveillance systems in place can detect increases in the rate at which people are becoming sick and alert public health officials that an outbreak event may be under way. Early indication of such an event can help a community minimize illness levels. Action plans for notifying high risk persons (the immuno-compromised, aged, infants, etc.) to cryptosporidiosis are needed. By taking additional protective measures, the public can also participate in protecting itself against the spread of disease.

Once alerted that a disease outbreak event may be occurring, epidemiologists attempt to identify patterns in the potential exposure sources of those persons known to be ill. In this manner, the source of *Cryptosporidium* can be implicated and measures can be implemented to eliminate additional exposure events. However, this is an imperfect science, and often no clear trends exist to identify the exposure source. Drinking water supplies are identified as the most likely source of

Table 2.1

Summary of waterborne cryptosporidiosis events

Location (dates)	*Cryptosporidium* found in water supply	Estimated number of people affected*	Notes on outbreak event	Reference
Braun Station, TX (May-July 1984)	Not measured	2,000	Sewage contaminated artesian well source of public drinking water.	D'Antonio et al., 1985.
Sheffield, England (May-June 1986)	Presence in raw water	935 (62)	Agricultural, nonpoint source pollution during heavy rainfall suspected. Unfiltered system.	Rush et al., 1990
Bernalillo County, NM (July-Oct 1986)	Not measured	(78)	Untreated surface water supply for drinking and recreational activities. Surface runoff from livestock grazing areas.	CDC, 1986.
Carrollton, GA (Jan-Feb 1987)	Presence in finished water	13,000	Sewage contamination at intake and filter start-up without backwashing.	Hayes et al., 1989.
Ayrshire, Scotland (Apr 1988)	Presence in distribution system	(27)	Cross-connection to sewage contaminated source in a distribution system tank.	Smith et al., 1989
Swindon and Oxfordshire, England (Jan-Mar 1989)	Presence in raw and finished water	(516)	Suspected agricultural, nonpoint source pollution during heavy rainfall event period.	Richardson et al., 1991
Berks County, PA (Aug 1991)	Found in raw well water supply	551	Chlorination only provided with possible septic tank and creek infiltration.	CDC, 1993.

(continued)

Table 2.1 (continued)

Location (dates)	*Cryptosporidium* found in water supply	Estimated number of people affected*	Notes on outbreak event	Reference
Jackson County, OR (Jan-June 1992)	Found in raw spring supply and unconfirmed in raw river supply	15,000	Chlorination only provided for the spring supply. Drought conditions in river receiving wastewater discharge with inadequate treatment of the river supply.	Oregon Health Division, 1992
Milwaukee, WI (Jan-Apr 1993)	Not measured	403,000	Suspected influx of organisms during rainfall events and treatment effectiveness was compromised.	Fox and Lytle, 1996
Cook County, MN (Aug 1993)	Raw lake water tested positive but no detections in finished water	27 (5)	Backflow from toilet system and septic tank to raw water line suspected.	Minnesota Department of Health, 1996
Las Vegas, NV (Jan-Apr 1994)	Not detected in raw, finished, or distribution samples	(78)	Increased incidence of disease with no clear delineation of the exposure source.	Goldstein, 1995
Walla Walla County, WA (Aug-Oct 1994)	Detected in well water and wastewater supplies	86 (15)	Seepage into well of treated wastewater used for irrigation.	Solo-Gabriele and Neumeister, 1996
Alachua County, FL (July 1995)	Detected in the wastewater	(72)	Backflow from a wastewater line in a camp's drinking water system.	Solo-Gabriele and Neumeister, 1996
Collingwood, Canada (1996)	Presumptive presence only	Not Available	Unfiltered system using chlorine for disinfection. The exposure source is still unclear.	*Cryptosporidium* Capsule, 1996

*The estimated number of people affected is based on epidemiological estimates. The number of confirmed cases is shown in parentheses.

Cryptosporidium exposure when an elevated "risk" of illness has been found to be associated with drinking water consumption levels, but clear identification of the true exposure source rarely results. Analyses of some of the outbreaks identified to date, however, have found evidence of *Cryptosporidium* in the drinking water supply, likely sources of contamination to the supply, and/or inadequacies in the treatment provided for protection against *Cryptosporidium* — all of which increase the creditability of epidemiologic evidence.

Finally, water utilities can be proactively involved in identifying the potential for outbreak events in their service areas and should work together with their local public health officials and health care community. The New York City Department of Environmental Protection (DEP) convened a panel of experts to evaluate current health department disease surveillance programs (Frost et al., 1996). That panel recommended several activities which water utilities and public health officials may implement to improve disease tracking:

- Designate an individual who is specifically responsible for coordinating waterborne disease surveillance.
- Monitor visits to hospital emergency rooms for enteric disease.
- Monitor sales of prescription and nonprescription medications for diarrheal illness.
- Conduct special enteric disease surveillance studies of nursing home and retirement home populations.
- Conduct surveillance of managed health care populations.
- Conduct surveillance of high-risk populations.

What can a utility do to protect its consumers against *Cryptosporidium*?

The best advice for public water utilities in protecting consumers against *Cryptosporidium* exposure is to consider implementing as many as possible of the recommendations contained in the AWWA White Paper, *What Water Utilities Can Do to Minimize Public Exposure to* Cryptosporidium *in Drinking Water* (1995). The following are some of the key elements identified in the White Paper:

- Assure that the highest quality source water available to your system is used as your drinking water supply.
- Work with neighbors who have source waters within the same watershed to promote and implement source water protection measures that could reduce the overall levels of *Cryptosporidium* entering your systems.
- Optimize your existing facilities for treatment performance in removing particles and achieving the most suitable conditions for effective inactivation.
- Evaluate and implement changes in your operational practices that can improve the overall reliability of treatment performance day-in and day-out.
- Prepare for all possibilities. Develop an action plan that identifies the key decision makers within your utility and the roles they will play if an outbreak does occur in your community. Include in the action plan provisions for communicating with your local and state health departments, the media, the public, and your utility's stakeholders (e.g., council or board members).

Additionally, utilities may want to participate in AWWA's "Partnership for Safe Water" program to ensure that their drinking water is as protected as possible against microbial contamination, especially by *Cryptosporidium*.

How does a utility know if *Cryptosporidium* is in its drinking water supply, and are there any indicators that *Cryptosporidium* is present?

Repeated sampling and analysis by an experienced laboratory can give a utility an approximate range of *Cryptosporidium* contamination levels or at least an indication of the frequency at which *Cryptosporidium* is present in the utility's source waters. Because there is no definitive test for *Cryptosporidium*, every utility that uses surface water or groundwater under the direct influence of a surface supply should apply treatment as if *Cryptosporidium* is present. To date, the usefulness of surrogates or indicators of *Cryptosporidium* occurrence in source waters is still unclear. However, important research funded by the AWWA Research Foundation is currently under way to clarify this issue and identify those surrogates that can be used in estimating the vulnerability of supply sources to *Cryptosporidium* occurrence.

Is it possible to test for *Cryptosporidium* in drinking water and in people?

It is possible to test, but the test results are only as valid as the method used for analysis. Current analytical methods for *Cryptosporidium* in drinking water have many limitations (see the discussion in Chapter 4 for more insights into the monitoring and analytical methods for *Cryptosporidium* in drinking water). However, tests for *Cryptosporidium* in people can be quite reliable when a person is symptomatic. The main reason for this dichotomy is that when a person is infected with *Cryptosporidium*, large numbers of oocysts can be found in feces. For environmental samples, like drinking water sources, the concentration of oocysts is much lower, making detection and quantification more difficult.

A limitation in the testing for *Cryptosporidium* in people is that someone must actually be releasing or "shedding" oocysts in stool at the time of testing for *Cryptosporidium* to be detected. Many times, people may be infected but are nonsymptomatic and still can produce oocysts in stool. Others that are both infected and symptomatic may only shed oocysts for a limited period of time during the illness. For these reasons, an absence of oocysts in stool samples does not necessarily mean that a person does not have cryptosporidiosis. However, when large numbers of oocysts are found in stool samples, this is confirmation of being infected with *Cryptosporidium*.

Is *Cryptosporidium* currently regulated for U.S. drinking water supplies?

While the EPA does currently regulate the quality of surface water supplies for pathogenic organisms, *Cryptosporidium* was *not* targeted directly in the Surface Water Treatment Rule (SWTR) as promulgated in 1989. That is not to say that the SWTR is necessarily insufficient for controlling for *Cryptosporidium* in drinking water supplies. However, analyses of the more recent cryptosporidiosis outbreaks associated with waterborne pathways have indicated that improvements to the SWTR may be appropriate for the control of *Cryptosporidium*. EPA and various stakeholders, including AWWA and AWWA Research Foundation, are involved in the evaluation of alternative control strategies to improve public health protection. Interim measures are expected by the end of 1998 (Interim Enhanced Surface Water Treatment Rule) with long-term control strategies implemented after the year 2000.

CHAPTER 3
OCCURRENCE IN SOURCE WATERS

Cryptosporidium occurs in source waters used for drinking water supplies. While *Cryptosporidium* does not multiply in the environment, the oocyst form of the organism is very resistant to many extremes in environmental conditions. The persistence of the organism contributes to its threat to drinking water sources. The following discussion highlights some of those concerns and what trends in source water occurrence utilities might expect to find.

What are the expected levels and patterns of *Cryptosporidium* in U.S. drinking water supply sources?

Numerous researchers and utilities have monitored for *Cryptosporidium* in source waters, demonstrating that a wide range of concentrations may occur and showing a high rate of detection in environmental samples (Table 3.1). While concerns regarding the accuracy of the analytical methods need to be considered in interpreting this information, *Cryptosporidium* has been clearly demonstrated to occur in U.S. source waters.

Considerable debate is still under way as to the nature of *Cryptosporidium* occurrence in source waters. The temporal variation of *Cryptosporidium* occurrence may behave similar to rainfall events, for example. Sometimes it is there, and sometimes not. For some source waters, especially those that receive wastewater discharges, researchers believe that oocysts are present all the time, but the concentration varies due to numerous factors. Understanding the expected levels and patterns of *Cryptosporidium* occurrence in source waters is critical to identifying the treatment challenges a utility faces in order to produce a safe drinking water supply. AWWARF is conducting a number of studies to investigate this issue, and EPA is taking several steps to characterize the occurrence of *Cryptosporidium* in U.S. source waters. Specifically, EPA has required all public water systems serving more than 100,000 people to conduct 18 months of source water monitoring for

Table 3.1

Summary of *Cryptosporidium* occurrence in U.S. surface waters

Source type	Number of samples (n)	Positive samples (%)	Range of oocyst concentration (oocysts/L)	Geometric mean concentration (oocysts/L)	Reference*
Stream/River†	6	100	0.8-5,800	1,920‡	E
Stream	19	73.7	0-240	1.09	C
Stream/River	58	77.6	0.04-18	0.94	A
Stream/River	38	73.7	<0.001-44	0.66	G
River†	11	100	2-112	25§	B
River/Lake	85	87.1	0.07-484	2.70	H
River	22	31.8	0.01-75.7	0.58	I
River/Lake†	18	NA	7.1-28.5	17.8	D
Lake	20	70.7	0-22	0.58	C
Lake/Reservoir	32	75.0	1.1-8.9	0.91	A
Lake	24	58.3	<0.001-3.8	1.03	G
Lake	44	27.3	0.11-251.7	4.74	I
Pristine River	3	NA	NA	0.08	D
Pristine River	59	32.2	NA	0.29	G
Pristine Lake	34	52.9	NA	0.093	G
Pristine Spring	7	28.6	<0.003-0.13	0.04	G
Pristine Lake	11	9.1	0-0.003	0.003	F

Source: Lisle and Rose, 1995. Reprinted with permission.

NA=Information not available.

*References cited are coded as follows: A = Ongerth and Stibbs, 1987; B = Rose et al., 1988a; C = Rose et al., 1988b; D = Madore et al., 1987; E = Roach et al., 1993; F = Rose et al., 1991; G = LeChevallier et al., 1991a; H = National *Cryptosporidium* Survey Group, 1992; I = LeChevallier et al., 1991b.

†Impacted by domestic and/or agricultural waste.

‡Arithmetic mean.

§Data adjusted for recovery efficiencies.

Cryptosporidium. In addition, EPA is performing a supplemental survey of *Cryptosporidium* occurrence in surface waters for over 100 systems in the United States that serve more than 10,000 people.

Are there different kinds of *Cryptosporidium* oocysts in water, and what is their significance to water utilities?

Cryptosporidium oocysts can originate from a number of animal hosts, including cattle, swine, horses, deer, chicken, ducks, fish, turtles, guinea pigs, cats, and dogs. Those of concern to drinking water utilities are the oocysts derived from mammalian hosts (Table 3.2) that remain viable in the source water. As mammals, humans can be infected by oocysts that are infectious to other mammals. The species of *Cryptosporidium* known to be infectious to humans is *Cryptosporidium parvum*. Other species do exist and have been identified in a wide range of animal hosts, but they have not been shown to be infectious to humans. Recent research has, however, indicated that nonmammalian species, such as geese, ducks and seagulls, can be carriers of oocysts infectious to humans when exposed to oocysts derived from a mammalian host (Fayer et al., 1997). This finding implicates a greater degree of vulnerability than previously thought for the transmission of infectious oocysts by source waters and uncovered finished water reservoirs that can be contaminated by waterfowl.

Table 3.2

Sources of *Cryptosporidium* in mammals

Animal Source	Reference
Cattle	Fayer, 1996; Fayer et al., 1989
Pigs and horses	Fayer, 1996
Sheep, goats, deer, mice, rats, rabbits, cats, dogs, and guinea pigs	Fayer and Ungar, 1986
Monkeys	Badenoch, 1995

The identification of different types of oocysts by originating host is still under way. Current analytical methods do not indicate if an oocyst found in a water source originated from a mammal — which would be of concern to a drinking water utility — or from the nonmammalian animals that populate the source water, which would present no concern to human health.

Environmental stresses can affect oocyst viability during their transport from an animal host to a drinking water supply. For example, a substantial decrease in oocyst survival occurred when they were subjected to extremes in temperature, but at the temperature ranges of most ambient waters, no detrimental effects have been observed. Additionally, pH conditions across the range typically found in the treatment and distribution of public water supplies showed no direct effect on oocysts (Carrington and Ransome, 1994). Their survival in ambient waters has been shown to be more than 6 months (Roberston et al., 1992).

Can a water utility reasonably describe the vulnerability of its source waters to *Cryptosporidium* occurrence?

Characterizing the vulnerability of drinking water supply sources to *Cryptosporidium* contamination is one means of assessing the possible levels of *Cryptosporidium* occurrence in source waters. Vulnerability components include: (1) potential contamination by specific land use activities or point source discharges to surface water supplies; (2) species and pathogenicity of the oocysts introduced by the various sources of contamination within a watershed; (3) environmental fate of the oocysts; and (4) the availability of surrogate parameters to indicate the presence and/or concentration of *Cryptosporidium* oocysts.

Substantial differences have been shown in *Cryptosporidium* oocyst levels between protected and unprotected portions of watersheds (Tables 3.3 and 3.4). Qualitatively, these results demonstrate how even minor levels of human activity can introduce oocysts into a water supply rendering it somewhat more vulnerable. When potentially large sources of *Cryptosporidium,* such as dairy farming, are present in a watershed, substantially higher levels of oocysts were found, indicating an even greater degree of vulnerability due to the presence of high-loading sources of *Cryptosporidium*. Land use activity is one of the most important elements of defining watershed vulnerability to

Table 3.3

Watershed land use impacts on *Cryptosporidium* occurrence

Sampling Site Description	Samples (n)	Oocysts Present (%)	Concentration Range* (oocysts/L)	Geometric Mean Concentration* (oocysts/L)
Protected drinking water supply	6	81	0.15-0.42	0.24
Pristine river, forestry area	6	100	0.46-6.97	1.62
River below rural community in forested area	6	100	0.54-3.6	1.07
River below dairy farming agricultural activities	6	100	3.3-63.5	10.72

Source: Hansen and Ongerth, 1991. Reprinted with permission.

*Results for positive samples only. Analytical method used was the IFA microscopic analysis with grab sample collection (no filtration step).

Table 3.4

Surface water monitoring results and detection limits

for an intensive sampling program in a Canadian watershed

Source description*	Samples (n)	Positive samples (%)	Range of oocysts (oocysts/L)	Geometric mean concentration (oocysts/L)	Detection limit range (oocyst/L)
Water Treatment Plant intake	27	52	0.017-0.443	0.035	0.006-0.57
Upstream of cattle	32	47	0.005-0.344	0.056	0.003-0.49
Downstream of cattle	30	47	0.014-3.00	0.133	0.006-1.50

Source: Ong et al., 1996. Reprinted with permission.

*Black Mountain Irrigation District watershed results only are presented.

oocysts — unfortunately, affecting changes in land use activities is one of the most difficult aspects of increasing watershed protection programs.

While land use activity clearly plays an important role in defining watershed vulnerability, relying on *Cryptosporidium* measurements alone to determine these impacts may be misleading. For example, in the study summarized in Table 3.4 cattle ranching activity increased the maximum concentration of *Cryptosporidium* oocysts by an order-of-magnitude in the watershed. However, the frequency at which oocysts were detected was identical at upstream and downstream locations to the cattle ranching activity (Table 3.4). Further, the range of detection levels for the up- and downstream locations also differed by the same order of magnitude as the maximum oocyst levels, confounding interpretation of both presence/absence frequencies and geometric mean estimates.

Given the difficulties in quantitatively assessing *Cryptosporidium* occurrence information, vulnerability characterization of watersheds may incorporate other water quality parameters to provide more insight into potential occurrence events over time. To date, no clear trends have been identified for such commonly used parameters as turbidity and coliform bacteria. Studies are currently under way to better identify appropriate surrogate or indicator parameters for use in source water vulnerability classifications.

What can utilities do to reduce *Cryptosporidium* oocysts from nonpoint sources of pollution?

Utilities do have some recourse in protecting their watersheds; however, the available measures are neither easy nor guaranteed for success. Local and state officials responsible for watershed protection provisions, whether from the perspective of natural resource conservation or water quality improvement, can be helpful in developing a dialogue with the agricultural community affecting a supply source. While prevention of *Cryptosporidium* being introduced into a source of supply may not be feasible, knowing the potential nonpoint sources of *Cryptosporidium* and their responses to hydrologic or climatological events can improve a utility's ability to protect its consumers. Being aware of the risks and indicators of possible events within the utility's watershed can form the basis of proactively implementing appropriate treatment responses.

What is the effect of wastewater discharges on *Cryptosporidium* levels in drinking water sources?

Wastewater discharges can be expected to have high concentrations of *Cryptosporidium* present on a consistent basis. The processes employed by wastewater treatment facilities will not inactivate oocysts but may physically remove some portion of them. Water utilities that have supply sources that are influenced by wastewater discharges should consider themselves highly vulnerable to *Cryptosporidium* occurrence.

A study of raw sewage (Madore et al., 1987) found between 870 and 5,280 oocysts per liter, but at one site receiving waste from a slaughterhouse, almost 14,000 oocysts per liter were found. Sewage effluent levels of *Cryptosporidium* were also found to be elevated with oocysts present in all nine sites sampled with concentrations ranging from 140 to 3,960 oocysts/liter. In contrast, Parker et al. (1993) collected 70 samples from seven wastewater treatment facilities (treated wastewater effluent) and found positive result in 37% of the samples with oocyst levels ranging from 0.03 to 2.4 oocysts/liter. These few studies illustrate the potential range in oocyst concentrations contributed by domestic wastewater discharges to receiving streams.

CHAPTER 4
MONITORING AND ANALYSIS

In protecting public health, one of the most valuable tools available to utilities is the monitoring and analysis of their supplies for potentially harmful constituents. Characterizing the occurrence of *Cryptosporidium* in drinking water supplies is a challenge for water utilities. Managers, engineers, operators, and the public all need to understand what monitoring for *Cryptosporidium* can indicate about drinking water supplies.

Should we monitor for *Cryptosporidium?*

The limitations of the available standardized analytical methods also limit the usefulness of *Cryptosporidium* monitoring results.

Cryptosporidium monitoring *can* indicate (1) the presence of oocysts in a drinking water supply, (2) a potential concentration range at which they may occur over time, and (3) when sufficient numbers of samples are collected over time, relationships with other attributes of a supply source (i.e., rainfall events, seasonal variations, and agricultural activity levels).

As of this writing, monitoring for *Cryptosporidium cannot* indicate (1) an absence of oocysts from a water supply, (2) the "true" concentration present in water for any particular sample, and (3) whether the oocysts present are still alive (i.e., "viable"), nor whether those oocysts would be infectious to humans. Nor can monitoring afford protection from *Cryptosporidium* occurrence.

Additionally, monitoring results may contain false positives or false negatives. False positives occur when other particles present in a sample behave and look like oocysts to the analyst. False negatives occur when oocysts do not behave or look like they should. Each sample collected for *Cryptosporidium* analysis must be searched for colorless particles that are only 4-6 μm in diameter and have the general characteristics of oocysts. The search takes place amidst numerous other waterborne particulates. This challenging task requires extensive skill and experience with microscopic examination but is also constrained by elements in the method beyond the analyst's control.

Is there a standard method for monitoring?

A standard *Cryptosporidium* detection method is included in *Standard Methods for the Examination of Water and Wastewater* (APHA, AWWA, and WEF 1994). However, this method excludes an important microscopic step that helps to keep other organisms from being misidentified as *Cryptosporidium*. The method included in the U.S. EPA's Information Collection Rule (ICR) (1996) is the procedure most commonly referenced in the United States (see Figure 4.1). This method is referred to as the ICR method and includes three steps: sample collection, concentration, and identification. Both techniques rely on the use of immunofluorescence assay (IFA) techniques whereby cleaned-up samples are stained using a fluorescent tag specific to *Cryptosporidium* oocysts and then examined microscopically to identify oocysts.

Scientists hope to have a new standard method or methods in the near future for improved analysis of *Cryptosporidium*.

What are the advantages and disadvantages of the method developed for the Information Collection Rule?

The advantage of the ICR method is that it has been used to find *Cryptosporidium* in raw and finished water sources, which has made possible increased efforts by the water industry to prevent its passage into drinking water. These efforts have included the initiation of improvements for numerous water treatment systems.

The disadvantages of the method are that inaccurate and imprecise results can be generated (Table 4.1 and 4.2). *Cryptosporidium* may be reported as absent when it is present or present when it is absent. In addition, types of *Cryptosporidium* that will not infect people may be reported, and there is no way to decide if oocysts are dead or alive with this method.

The importance of analytical error in interpretation of *Cryptosporidium* monitoring results should also be understood. The ability to recover the oocysts present in a sample through the sample

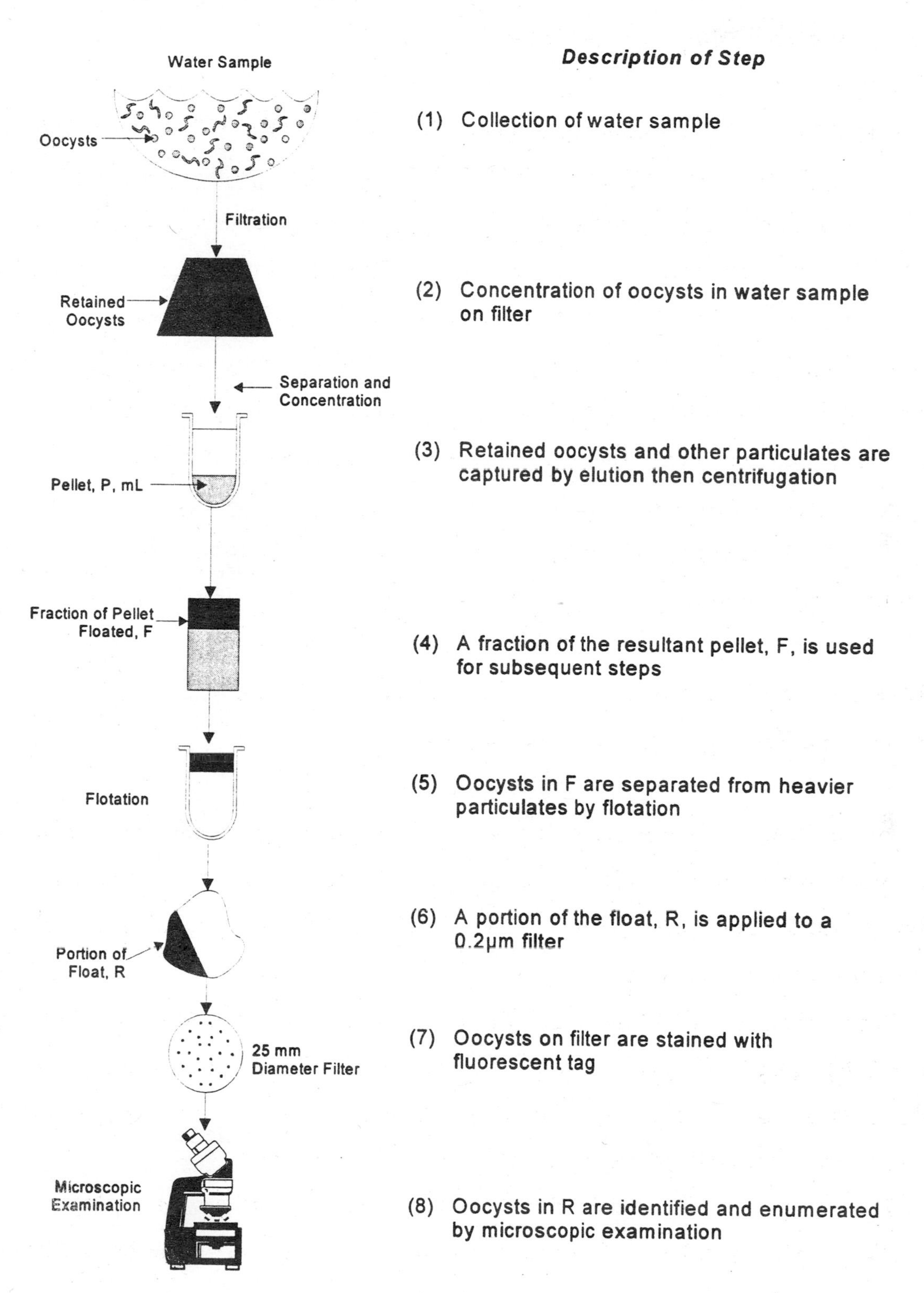

Figure 4.1 Analytical method for *Cryptosporidium* as described in the Information Collection Rule

Table 4.1

Summary of laboratory performance for protozoan measurement accuracy and precision

Study reference (number of labs participating)	*Giardia* mean recovery/SD*/CV† (%)	*Cryptosporidium* mean recovery/SD*/CV† (%)	False sample results (positive/negative)
Clancy et al., 1994 (13)	5.5/7.9/144	1.1/1.7/155	yes/yes
Clancy et al., 1995 (10)‡	44/18/58	23/16/111	yes/yes
Clancy et al., 1995 (13)‡	25/25/93	35/36/105	yes/yes
Clancy, 1996a (8)	23/10/43	5/5/100	yes/yes

*SD is the standard deviation of the mean recovery found for participating laboratories.

†CV is the coefficient of variation, defined as the ratio of the standard deviation to the mean.

‡Two independent rounds of performance evaluation testing were conducted and are reported separately in this table.

Table 4.2

Summary of field spiking study results

Protozoan organism	Mean recovery/SD/CV (%)	Range of recovery (%)
	Raw water samples (n = 167)	
Giardia	36.7/29.6/94.7	0.7-170.8
Cryptosporidium	11.0/12.7/115.4	0.0-82.9
	Finished water samples (n = 112)	
Giardia	26.4/25.0/154.3	0.4-154.3
Cryptosporidium	7.2/8.7/120.5	0.0-46.6

Source: Clancy 1996b

collection, handling, processing, and analysis is an important feature of the accuracy of the ICR method. The results from recent multiple laboratory testing studies have shown that low recoveries result from the ICR method and a high degree of variability in laboratory performance also can be found (Table 4.1). Further, false positive and negative results were reported by many of the laboratories participating in these studies.

The results presented in Table 4.1 show the ability of laboratories to measure *Cryptosporidium* under ideal conditions since these were performance evaluation type samples. However, utilities should be most concerned with the ability to measure *Cryptosporidium* in the source and finished waters. Field testing results (Table 4.2) indicate once more that low recoveries of *Cryptosporidium* are to be expected with the ICR method (11% in raw water samples and 7.2% in finished water samples). The variation in recovery rates was also large, demonstrating that inconsistent results can be expected in *Cryptosporidium* measurements.

Are there any new methods on the horizon for utility use?

Several new techniques are in the various stages of research development in the United States, United Kingdom and Australia. None have yet been adopted as a standard method or tested in several laboratories. We expect to see some new methods ready for routine laboratory use in the near future. An exhaustive description of the newer methods and method components can be found in Frey et al. (1997). Until the accuracy and precision of new techniques have been demonstrated, the user should be aware of any limitations or uncertainties associated with using unproven methods.

What analytical method should be used during the next couple of years?

Use laboratory techniques that have demonstrated the highest and most consistent recovery values so as to have greater confidence in monitoring results. Candidate methods should document the following to assure their performance: a standard operating procedure that has been challenge-tested; recovery efficiency in high and low turbidity water containing various types of organic and inorganic particulates; the type of oocysts used for the challenge tests; and the techniques used to count the oocysts in the challenge test.

What is the best way to select a laboratory?

A list of laboratories approved for ICR analyses may be obtained from the U.S. EPA Laboratory Coordinator for the Information Collection Rule or the SDWA Hotline (800-462-4741). The AWWA Information Collection Rule A-Team also can provide the list of approved laboratories for *Cryptosporidium* analysis (800-200-0984; or at the following e-mail address 103327.2057@compuserve.com).

If you would like to know more about the performance of laboratories before making your selection (even from the list of approved laboratories for ICR monitoring), here are some information requests that you can make:

- documentation of their education, experience, and round-robin performance
- in-house quality control and assurance data and the frequency in which they are generated
- written records of the laboratory's standard operating procedure (preferably, this is a method recommended by more than one laboratory or published in a peer-reviewed journal)

What are the analytical terms for the presence of *Cryptosporidium* as reported by laboratories using the ICR method?

Analytical laboratories performing *Cryptosporidium* analyses will report several types of results that may appear confusing. When a sample is analyzed for *Cryptosporidium*, a microscopic examination is performed and the analyst counts the number of organisms present. Therefore, the results reported by laboratories will represent the "counts" found in a water sample. Four types of results may be reported by a laboratory:

- *Total IFA counts* include *Cryptosporidium*-sized and *Cryptosporidium*-shaped objects that react positively with fluorescent stains and have characteristics typical of *Cryptosporidium* oocysts. Because there are known cross-reactions of organisms (e.g., organisms that react like *Cryptosporidium* to the immunofluorescent antibody), the possibility of incorrect identification in both raw and finished water exists. Microscopists can lower the risk of inaccurate identification by becoming familiar with cross-reacting organisms.
- *Empty counts* are a subset of total IFA counts in which the fluorescing objects have no internal features. These counts represent the number of organisms that could have been *Cryptosporidium* oocysts that are no longer viable but for which the shells remain intact.
- *Counts with amorphous internal structures* are a second subset of total IFA counts in which the fluorescing object has featureless material present within the cell. These counts represent the number of organisms that could be *Cryptosporidium* oocysts and that may still be viable.
- *Counts with internal structures* are the last subset of total IFA counts in which the distinctive organelles for *Cryptosporidium* were seen. These counts represent the number of organisms that are confirmed as *Cryptosporidium* oocysts and that may still be viable.

None of the counts presented in the preceding list reflects the viability or infectivity of oocysts; rather they imply whether an oocyst is whole (internal structures present) or not (empty), and they indicate confidence that the organism observed under microscopic examination is really a *Cryptosporidium* oocyst.

What do the analytical results mean?

The reported absence of *Cryptosporidium* may indicate that no oocysts are present. However, it can also mean that oocysts were in the sample but were not detected due to limitations in the analytical methods. In the ICR method, the absence of *Cryptosporidium* is reported as a detection limit for the measurement of organisms in the sample. This is not a true detection limit because the efficiency of the method has not been included.

When oocysts have been detected or counted, the results presented depend on the type of count found (as defined earlier). When only total counts are provided, a utility cannot be sure whether the counts represent confirmed oocysts or even whether a whole organism was present. Utilities should request that their laboratories report all four types of counts: total, internal structures, amorphous internal structures, and empty. The total count represents the potential maximum level of *Cryptosporidium* detected in a sample and provides a conservative estimate of the vulnerability of supplies to oocysts. Internal structures represent confirmed *Cryptosporidium* in a sample and should be of primary concern to a utility. The amorphous internal structure counts represent possible *Cryptosporidium* oocysts that could have internal structures which were not apparent to the microscopist. Empty counts indicate that oocysts may have been present in a supply, but these do not represent a health risk.

Finally, a laboratory may estimate the concentration of oocysts in a sample based on the number of organisms counted and volume analyzed. For example, if a laboratory counted 2 oocysts in a volume of 0.1 liter of sample that was analyzed, the estimated concentration would be 20 organisms per liter of water. This estimated concentration may then be further adjusted by the laboratory based on its recovery rate of oocysts (the fraction of oocysts found from a known concentration condition). Ask the laboratory to document how the concentration of oocysts in the reported results was determined. Using a laboratory recovery rate to project concentrations present may not be appropriate.

CHAPTER 5
TREATMENT EFFECTIVENESS: REMOVAL AND INACTIVATION

An effective barrier to *Cryptosporidium* entering consumers' drinking water supplies is the physical and chemical treatment of water supplies to remove and inactivate *Cryptosporidium* oocysts. To understand the difference between removal and inactivation, consider the terms used in describing analytical results for *Cryptosporidium*. On the one hand, removal of oocysts by treatment processes results in lower numbers of total counts in the treated water than in the raw water supply. Inactivation, on the other hand, results in higher numbers of empty counts (i.e., empty shells only of the oocysts are found) after treatment than in the raw water supply but the number of total counts may remain the same. Thus, treatment processes for oocyst removal physically eliminate the organisms during treatment, while inactivation processes "kill" the organism, potentially leaving behind its empty shell.

The majority of water utilities in the United States that have surface water sources employ some form of physical removal treatment, and all surface water supplies practice disinfection for the inactivation of pathogens. The questions being raised concern to what extent these treatment processes are effective in controlling *Cryptosporidium*.

Which water treatment processes provide removal of *Cryptosporidium* and which are responsible for inactivation?

Removal of *Cryptosporidium* oocysts can be achieved to some extent by any water treatment process that captures and removes particles from the water supply. The most common form of particle removal, and therefore *Cryptosporidium* removal, at water treatment plants is coagulation with or without settling followed by granular media filtration. Membrane technologies also remove particles and can eliminate *Cryptosporidium* oocysts present in water supplies. Even some water treatment processes not intended for particle removal can be used to remove oocysts, such as ion exchange resins or activated carbon filters.

Inactivation of oocysts can be accomplished, to some extent, by adding an oxidant to water. Many disinfectants used to control pathogens in water supplies are also known as oxidants. Some chemicals commonly used in water treatment are also oxidants, but their use has only a negligible effect on oocysts (e.g., potassium permanganate). While the majority of water treatment facilities in the United States use chemicals for inactivating pathogens, energy-based processes can also be used. A growing number of wastewater and small water treatment facilities are employing ultraviolet radiation (UV) for the inactivation of pathogens, but this technique may not be suitable for *Cryptosporidium* inactivation. Newer technologies are also under development.

How is the effectiveness of water treatment processes evaluated for *Cryptosporidium* control?

The effectiveness of removal and inactivation processes for *Cryptosporidium* is evaluated using "challenge" test conditions in bench- and pilot-scale applications. To test the removal efficiency of a process, large numbers of oocysts of a known quantity are added to the raw water so that following treatment, the remaining oocysts can be measured. The difference between the raw and finished water concentrations represents the removal efficiency of the treatment process and can be expressed either in terms of percent removal (%) or log removal (logs) as shown in the following:

Percent Removal Efficiency ($\eta_{\%}$):

$$\eta\% = \frac{(\text{Raw Total Oocysts} - \text{Finished Total Oocysts}) \times 100\%}{\text{Raw Total Oocysts}}$$

Log Removal Efficiency ($\eta_{\log}$):

$$\eta_{\log} = \text{Log}_{10}\,(\text{Raw Total Oocysts}) - \text{Log}_{10}\,(\text{Finished Total Oocysts})$$

These measures of removal efficiency are the same used for inactivation, but the measurements used for oocysts are different. In removal studies, the total number of oocysts in the raw and finished water samples are used to determine the percent or log removal efficiency. For inactivation testing, the number of viable (alive) oocysts are measured for the raw and finished water to determine the percent or log inactivation efficiency, as shown in the following:

Percent Inactivation Efficiency ($\eta_{\%}$) :

$$\eta_{\%} = \frac{(\text{Raw Viable (Infectious) Oocysts} - \text{Finished Viable (Infectious) Oocysts}) \times 100\%}{\text{Raw Viable (Infectious) Oocysts}}$$

Log Inactivation Efficiency (η_{log}) :

$$\eta_{log} = \text{Log}_{10}(\text{Raw Viable Oocysts}) - \text{Log}_{10}(\text{Finished Viable Oocysts})$$

The performance of a water treatment process for *Cryptosporidium* control must be challenged to the degree of its ability to remove or inactivate the oocysts. For example, if a technology can remove 10 oocysts, adding only 9 oocysts to the raw water to "challenge" the technology will not generate conclusive evidence that the technology could remove 10 organisms. Therefore, challenge tests are typically designed to capture several orders of magnitude of treatment performance so that the full range of a technology's performance can be measured. For convenience, the use of the log removal or inactivation efficiency terminology has become commonplace in describing treatment performance since it provides a shorthand to the multiple orders-of-magnitude used in evaluating treatment performance (Table 5.1).

What water treatment processes are effective in removing *Cryptosporidium*?

All physical/chemical treatment processes, such as coagulation, sedimentation, dissolved air flotation, and filtration and membrane technologies, will remove *Cryptosporidium* to some degree. The effectiveness of such barriers differs depending on the process or combination of processes.

Table 5.1

Example of *Cryptosporidium* oocyst challenge-test results expressed in terminology usually used for evaluating the removal and inactivation performance of treatment technologies

Raw water *Cryptosporidium* level (oocysts/L)	Finished water *Cryptosporidium* level (oocysts/L)	Percent removal/inactivation efficiency (%)	Log removal/inactivation efficiency (logs)
10	5	50	0.3
	1	90	1.0
100	25	75	0.6
	10	90	1.0
	5	95	1.3
	1	99	2.0
1,000	100	90	1.0
	25	97.5	1.6
	10	99	2.0
	5	99.5	2.3
	1	99.9	3.0
10,000	1,000	90	1.0
	500	95	1.3
	100	99	2.0
	25	99.75	2.6
	10	99.9	3.0
	5	99.95	3.3
	1	99.99	4.0

Table 5.2 summarizes the results of available bench- and pilot-scale data for the physical removal of *Cryptosporidium* by water treatment processes. While this synthesis of the literature may not be complete, some clear trends can be discerned from those studies evaluated:

- The variability in treatment performance found from controlled bench- and pilot-scale tests was much less than that observed at full-scale treatment plants.
- Coagulation with gravity settling alone can achieve up to approximately 1.5-logs of *Cryptosporidium* removal.
- When coagulation and filtration are practiced without sedimentation (direct filtration), the removal efficiency is about 2.0 to 4.0-logs.

Table 5.2

Summary of the *Cryptosporidium* removal efficiencies estimated for various physical and chemical processes

Treatment process description	Removal achieved at testing scale (log_{10})		
	Bench scale	Pilot scale	Full scale
• Coagulation + Gravity Settling	<1.0[A]	1.4-1.8[B]	0.4-1.7[G]
• Coagulation + Filtration		2.7-5.9[B]	1.6-4.0[E]
		2.5-3.8[H]	
		2.7-2.9[I] *	
• Coagulation + Gravity Settling + Filtration		4.2-5.2[B]	1.6-4.0[E]
		>5.3[F]	<0.5-3.0[F]
		2.1-2.8[I] *	1.0-2.5[G]
• Coagulation + Dissolved Air Flotation (DAF)	2.0-2.6[A]		
• Coagulation + DAF + Filtration†			
• Slow Sand Filtration		>3.7[C]	
• Diatomaceous Earth (DE) Filtration		>4.0[C]	
• Coagulation + Microfiltration		>6.0[D]	
• Ultrafiltration		>6.0[D]	
• Nanofiltration†			

Sources: References are encoded as follows: A = Plummer et al., 1995; B = Patania et al., 1995; C = Schuler et al., 1988; D = Jacangelo et al., 1995; E = Nieminski and Ongerth, 1995; F = LeChevallier et al., 1991b; G = Kelley et al., 1994; H = Anderson et al., 1996; and I = Nieminski, 1995.

*Range of average removal efficiencies based on reservoir and river water sources.

†Published results are not available to document the removal performance of this technology to any testing scale.

- For conventional treatment (coagulation, settling, and filtration), *Cryptosporidium* removal efficiencies ranged from approximately 2-logs to over 5-logs.
- Membrane technologies including micro-, ultra-, and nanofiltration can reliably achieve more than 6-logs of *Cryptosporidium* removal.

The reliability of treatment is also an important issue. While all treatment processes can have "failure" events, normal operational variability can impose a range of performance for any technology. Reliability addresses the resistance of a technology to those types of changes so that a consistently safe drinking water supply is produced.

What is the inactivation performance of disinfectants for *Cryptosporidium*?

Cryptosporidium oocysts have been shown to be resistant to the conventional disinfectants used in water supplies, free and combined chlorine. In some studies, a tenfold increase in contact time was required to inactivate a comparable number of *Cryptosporidium* oocysts as compared to *Giardia* cysts. Advanced oxidation techniques using ozone and chlorine dioxide have also been investigated to identify their performance in inactivating oocysts. While these disinfectants are more effective than chlorine and chloramines for oocyst kill, *Cryptosporidium* is still much more difficult to inactivate than other pathogens commonly encountered in water supplies.

Preliminary evidence of synergistic effects from combinations of inactivation agents — where the performance of the combination of disinfectants has been greater than that of the individuals alone — has been noted by Finch et al. (1994), and continued research efforts are still in progress to further define these effects. The effects of synergism could demonstrate that disinfection schemes using multiple disinfectants — such as chlorine or ozone followed by chloramines — would be much more effective and feasible for water utilities than relying on a single disinfectant alone to generate sufficient inactivation performance to protect public health.

The effectiveness of disinfectants for *Cryptosporidium* inactivation differs by the strength of the oxidant following the trend shown below in inactivation performance (Table 5.3):

Ozone >> Chlorine Dioxide >> (Chlorine or Chloramines)

Unlike the effect of chloramines with *Giardia* inactivation, chloramines may be as or more effective in *Cryptosporidium* inactivation than free chlorine depending on pH. Further research is under way to more fully develop appropriate measures of inactivation by individual disinfectants and combinations of disinfectants used in water treatment plants. Some of the important aspects of disinfection practices being considered are (1) natural water matrix effects on disinfection performance, (2) disinfection kinetics and the mechanism of measuring CT values, (3) measurement techniques for inactivation (animal infectivity versus excystation versus viability staining), (4) response of environmentally stressed oocysts to inactivation, and (5) the sequence of disinfectant addition.

Table 5.3

Selected *Cryptosporidium* inactivation results for various disinfectants

Disinfectant type and reference*	CT needed (mg·min/L) for inactivation performance			
	68% (0.5-log)	90% (1-log)	99% (2-log)	99.9% (3-log)
Ozone				
Korich et al., 1990	1.75	3.00	3.75	4.50
Owens et al., 1994	—	—	5.5	—
Finch et al., 1994a	—	—	2.4†	3.7†
Chlorine Dioxide				
Korich et al., 1990	32.5	52.0	> 78.0	—
Free Chlorine				
Korich et al., 1990	3,600	4,800	7,200	8,400
Finch et al., 1994b	> 7,200	—	—	—
Chloramines				
Korich et al., 1990	3,200	5,600	> 9,600	—
Finch et al., 1994b	> 2400	—	7,200	—
Free Chlorine + Chloramines Finch et al., 1994b	—	60 Free Chlorine, 480 Chloramines‡	—	—
Ozone + Chloramines Finch et al., 1994a	—	—	6.2 Ozone, 480 Chloramines§	—

*Test conditions represented in the results presented are as follows: (1) Korich et al., 1990: pH 7.0, temperature 25°C, animal infectivity results; (2) Owens et al., 1994: pH 7.5-8.5, temperature 22-25°C, animal infectivity results; (3) Finch et al., 1994a: pH 6.9, temperature 22°C, animal infectivity results.

†CT values represent the nonlinear C^nT^m Hom-type model results. Conventional linear CT values for 2-log and 3-log *Cryptosporidium* inactivation were found to range from 1.7 to 8.1 and 3.1 to 15.3 mg·min/L, respectively, depending on the integrated ozone residual level (0.25 to 3.0 mg/L). For lower temperature conditions (7°C), the nonlinear CT values were 6.9 and 10.3 mg·min/L for 2- and 3-log inactivation of oocyst, respectively.

‡Combined disinfection by free chlorine and chloramines achieved 96.8% (1.5-log) inactivation of oocysts at 22°C.

§The CT value represented for ozone is the simple CT calculated as the ozone concentration multiplied by the contact time.

— indicates no data available

In addition to the disinfectants presented in Table 5.3, newer inactivation technologies are currently being investigated for their effectiveness toward *Cryptosporidium*. The AWWA Research Foundation has a study under way that is investigating electrochemical technologies for the inactivation of *Cryptosporidium*. While results are not yet available for this study, the opportunity for new methods of water treatment is clearly present.

How can a utility assess its routine plant performance for *Cryptosporidium* control?

Routine plant performance for *Cryptosporidium* control can be assessed through rigorous monitoring and tracking of particle removal and inactivation performance. Many plants can effectively use turbidity measures to determine particle removal, while others may need to rely on additional measures such as particle counting. Inactivation performance can be tracked by assessing the CT values achieved during routine operations. Trend analysis can support the development of appropriate alarm-response procedures so that reliable physical removal and inactivation performance take place. While monitoring for *Cryptosporidium* directly may assist in building reference occurrence information, reliance on those measurements to indicate the safety of the treated water is not advised. The available analytical methods cannot clearly distinguish when oocysts are absent. Therefore, surrogate or indicator measurements should be used to assess plant performance. For physical removal, indicators may include filtered turbidity, particle removal and/or filtered particle counts, and jar testing to determine optimal coagulant dose. Inactivation performance can be measured by the product of disinfectant residual concentration and contact time (CT). An integrated approach to disinfection operations is soon to be released from the AWWA Research Foundation.

Are drinking water supplies vulnerable to posttreatment contamination of *Cryptosporidium* either in the distribution system or within consumers' home plumbing?

All drinking water supplies are vulnerable to posttreatment contaminant of *Cryptosporidium* from such sources as cross-connections, incomplete disinfection following main repair or replacement, open finished water reservoirs, and inadequate household sanitation. Utilities can address those sources of contamination that originate with their practices or infrastructure. For example, reduced risk of contaminating the drinking water supply can be accomplished by implementing a rigorous cross-connection control program.

While many utilities already have standard operating procedures for the return to service of distribution system piping, these procedures may be insufficient for protection against *Cryptosporidium* contamination. Developing a practical method of assuring the sanitary integrity of distribution systems is an important task on the horizon for drinking water utilities. Covering finished water reservoirs in some cases is not feasible; however, posttreatment may be appropriate in some cases to reduce public health risks. Utilities can assist their consumers in protecting themselves through education and public outreach programs that address household sanitation and exposure to potential pathogens such as *Cryptosporidium*.

Will future drinking water standards include requirements for *Cryptosporidium* control?

EPA has made a commitment to the drinking water industry, to Congress and to the public that regulatory controls would be implemented for the protection of public health against *Cryptosporidium* in drinking water supplies. The Interim Enhanced Surface Water Treatment Rule (IESWTR) is the first step in that direction. The long-term microbial control strategy adopted in the final ESWTR will be substantially related to appropriate *Cryptosporidium* controls.

CHAPTER 6
PUBLIC INFORMATION AND MANAGEMENT

Water utilities throughout the United States will continue to be in the public "spotlight" with respect to *Cryptosporidium* issues. The safety of water supplies is coming under greater scrutiny, and maintaining public confidence through a potential crisis or outbreak event is an important aspect of managing water utilities. The utility membership of AWWA believed that this was such an important component of dealing with the issue of *Cryptosporidium* that in 1996 an assembly of experts and utility managers, using the American Assembly process, prepared a statement on "Managing for Water Quality and System Reliability." Here are some suggested steps contained in the AWWARF American Assembly Statement for the improvement of operational and management performance in addressing *Cryptosporidium*-related issues (AWWARF 1997).

Planning is the key to the prevention, mitigation, and remediation of potential crises due to *Cryptosporidium* in drinking water supplies

Strategic or master planning sets a path for management and staff to follow for the maintenance and improvement of water quality and operations in producing safe drinking water supplies. A strategic plan may include numerous components, including a mission statement; an evaluation of the business aspects of the utility and specific goals to be accomplished; an assessment of the functionality of facilities and supply sources for long-term operational objectives; and an analysis of the financial requirements for investment and maintenance of a utility's operation (capital improvement plans, annual and operating budgets, etc).

Contingency planning establishes the communication and decision-making authorities for actions that can reduce or mitigate the extent of a water quality event under emergency conditions. Contingency plans should identify the range of actions that may be available to managers and operational staff with the appropriate decision-makers identified for each level of "crisis" represented in an event. Such plans may be developed for source water, water treatment plant, and distribution system events.

Special attention should be paid to the communication with the public, public health officials, drinking water regulators, community leaders and the media during such periods so that information is disseminated in a timely and appropriate manner. For more information on contingency planning, Cryptosporidium *and Water: A Public Health Handbook* (Working Group on Waterborne Cryptosporidiosis, 1997) is an excellent resource and is available through the CDC and AWWA.

Communication planning provides the foundation of day-to-day operation and management. Thousands of decisions are made daily in water utilities that can affect the performance and reliability of drinking water quality. Setting a communication pathway that incorporates decision making tasks and information management can assist utilities in identifying the risks of water quality changes or operational changes and the necessary steps for addressing such changes before a problem or crisis condition arises.

The focus of water quality management programs needs to include system optimization and reliability maximization, and this may force changes in work habits and performance

The last decade of changes to the drinking water industry, as motivated by water quality issues, has been focused on whether new or more advanced technologies are needed. A shift in this paradigm is needed where water quality management programs should first focus on system optimization — "How well can my system perform?" and reliability maximization — "How consistently can my system meet that performance level?" The optimization process should identify those operational and management activities that promote the production of the highest quality drinking water. Reliability evaluations, however, must accompany the optimization process. Water quality, treatability, and utility operations are dynamic and interdependent. The reliability of producing high quality drinking water is a measure of that dynamic relationship, and reflects the "true" performance capability of drinking water facilities.

Changes in the work habits, performance, skill levels and expectations for water utility staffs are likely to occur when water quality management programs based on system optimization and reliability maximization are implemented. The tools used may be different, including automation, information management systems, and computer programs to provide trend analyses and projection as well as optimization routines. Not only will skill sets change for the workforce, but also the potential exists for the "culture" to change — with responsibility and accountability being shifted to appropriate organizational levels.

CHAPTER 7

HOW TO LEARN MORE ABOUT *CRYPTOSPORIDIUM*

A plethora of information is available about *Cryptosporidium*, and for those having access to the World Wide Web, electronic searches will result in numerous home pages and references housing invaluable information. Here are some highlights in the World Wide Web (WWW) locations and some suggested library additions that every water utility should consider acquiring.

World Wide Web (Internet) home pages

- AWWA Research Foundation: http://www.awwarf.com/crypto97.htm
- U.S. Environmental Protection Agency: http://www.epa.gov
- AWWA: http://www.awwa.org
- Centers for Disease Control: http://www.cdc.gov/ncidod/diseases/crypto/crypto.htm
- Water Quality Information Center (U.S. Department of Agriculture and the University of Maryland): http://www.nalusda.gov/wqic
- Kansas State University: http://www.personal.ksu.edu/~coccidia

Some important contacts for more information

- EPA SDWA Hotline, Washington, DC: 800-462-4741
- AWWA Information Collection Rule A-Team, Washington, DC: 800-200-0984
- CDC AIDS Hotline, Atlanta, Georgia: 800-342-AIDS
- Arkansas Department of Health, Sue Casteel, Program Coordinator for the Surface Water Treatment Rule: 501-661-2623
- *Cryptosporidium* Capsule, Newsletter on *Cryptosporidium* Issues. Subscription inquiries can be mailed to FS Publishing, 241 Sixth Avenue, Suite 7E, New York, NY, 10012. Phone: 212-439-7203; Fax 212-439-7231; e-mail: crypcap@fspubl.com.

Suggested readings

AWWARF (American Water Works Association, Research Foundation). 1997. AWWARF American Assembly Statement. In J. Cromwell, D. Owen, J. Dyksen, and E. Means. *Managerial Assessment of Water Quality and System Reliability.* Denver, Colo.: AWWA and AWWARF.

Badenoch, J. 1995. *Cryptosporidium* in Water Supplies. Report of the Group of Experts. London: Department of the Environment, Department of Health, Her Majesty's Stationery Office.

Bellamy, W.D., G.R. Finch, and C.N. Haas. 1997. *Integrated Disinfection Design Framework.* Denver, Colo.: AWWARF and AWWA.

Finch, G.R., E.K. Black, L. Gyürek, and M. Belosevic. 1994a. *Ozone Disinfection for* Giardia *and* Cryptosporidium. Denver, Colo.: AWWARF and AWWA.

Finch, G.R., E.K. Black, and L. Gyürek. 1994b. Ozone and Chlorine Inactivation of *Cryptosporidium*. In *Proc. of the Water Quality Technology Conference*. Denver, Colo.: AWWA.

Hancock, C., J.V. Ward, K.W. Hancock, P.T. Klonicki, and G.D. Sturbaum. 1996. *Microscopic Particulate Analysis (MPA) for Filtration Plant Optimization.* U.S. EPA No. 910-R-96-001. Manchester Environmental Laboratory. Washington, D.C.: USEPA.

Jakubowski, W., S. Boutros, W. Faber, R. Fayer, W. Ghiorse, M. LeChevallier, J. Rose, S. Schaub, A. Singh, and M. Stewart. 1996. Status of Environmental Methods for *Cryptosporidium. Jour. AWWA,* 88(9):107-121.

Juranek, D.D., D.G. Addiss, M.E. Bartlett, M.S. Arrowood, O.G. Colley, J.E. Kaplan, R. Perciasepe, J.R. Elder, S.E. Regli, and P.S. Berger. 1995. Cryptosporidiosis and Public Health: Workshop Report. *Jour. AWWA,* 87(9):69.

LeChevallier, M.W., W.D. Norton, R.G. Lee, and J.B. Rose. 1991. Giardia *and* Cryptosporidium *in Water Supplies*. Denver, Colo.: AWWARF and AWWA.

Lisle, J.T., and J.B. Rose. 1995. *Cryptosporidium* Contamination of Water in the U.S.A. and U.K.: A Mini-Review. *Jour. Water SRT-Aqua*, 44(3):103-117.

CDC. 1995. Assessing the Public Health Threat Associated With Waterborne Cryptosporidiosis: Report of a Workshop. *Morbidity and Mortality Weekly,* Vol 44(RR-6): June.

Working Group on Waterborne Cryptosporidiosis. 1997. Cryptosporidium *and Water: A Public Health Handbook.* Atlanta, Ga.: Council of State and Territorial Epidemiologists (also distributed through AWWA, Denver, Colo.).

APPENDIX A
AWWARF REPORTS
ON *CRYPTOSPORIDIUM*

Research Foundation Subscribers can order completed reports by writing to AWWARF, 6666 West Quincy Avenue, Denver, CO 80235; by calling (303) 347-6121; by sending a fax to (303) 730-0851; or by sending e-mail via Internet to "rfreports@awwarf.com." Others, including AWWA members, can purchase reports from AWWA Member Services by calling (800) 926-7337. ***Please note that you can only order reports for completed projects that have order numbers.*** *More information on specific projects can be obtained by contacting Deborah R. Brink, Director, Research Management Division, at (303) 347-6109.*

1997 Projects

Critical Review of Existing Data on Physical and Chemical Removal of *Cryptosporidium* in Drinking Water (RFP 437)

To critically assess past *Cryptosporidium* research results and conclusions relative to current knowledge and state of interpretation of experimental results, and to describe the effectiveness of water treatment processes for removing *Cryptosporidium*. Develop a platform for comparison of research results on the removal of *Cryptosporidium* and a model to predict removal efficiencies. $150,000.

Validation of Serology Method for the Detection of *Cryptosporidium* in Human Populations (RFP 438)

To test and validate *Cryptosporidium*-specific serologic assay(s) by determining their sensitivity and specificity with samples collected from human populations. To determine the duration of IgM, IgG, and IgA antibody responses after exposure to *Cryptosporidium*. $150,000.

Update the AWWARF Report on Experimental Methodologies for the Determination of Disinfection Effectiveness to Include *Cryptosporidium* Disinfection Protocols (RFP 439)

Document a common experimental protocol for *Cryptosporidium* disinfection studies and ensure that projects funded through AWWARF, as well as other agencies, develop comparable results. $25,000.

Impact of Sample Collection and Processing on *Cryptosporidium* Viability and Infectivity (RFP 443)

Determine and quantify the impacts of sample collection, concentration, and purification methods on the viability and infectivity of *Cryptosporidium* oocysts. $350,000.

Protocol for *Cryptosporidium* Risk Communication for Drinking Water Utilities (RFP 444)

Develop written protocols for implementing voluntary and mandated *Cryptosporidium* risk communication programs, using standard consumer marketing strategies and established risk communication techniques, and provide methods for utilities to measure the effectiveness of the programs. $150,000.

Applications Guidance for Particle Counting Technology at Drinking Water Utilities (RFP 456)

Develop guidance for particle counting applications, including instrumentation, installation, data acquisition and analysis, and quality assurance in drinking water treatment. $100,000.

Filter Operation Effects on Pathogen Passage (RFP 490)

Evaluate the effects of operational factors (e.g., headloss, turbidity, plant-flow changes, particle counts, and charge) on the passage of pathogens through filters during filter ripening periods, filter-run turbidity and particle spikes, and filter breakthrough events. In addition, investigate the contribution of pathogen loading from recycled backwash water. $400,000.

Molecular Mechanisms of *Cryptosporidium* and *Giardia* Inactivation Using Sequential Chemical Disinfection (RFP 492)

Evaluate the molecular mechanisms of protozoan inactivation by sequential chemical disinfection. Determine the molecular basis for synergistic inactivation and the targets of action through various sequential combinations of disinfectant challenges. $300,000.

Bromate Formation and Control During Ozonation of Low Bromide Waters (RFP 493)

Evaluate the formation and control of bromate in low bromide waters under ozone dosages capable of inactivating *Cryptosporidium*. $300,000.

The following projects were approved, but will not have RFPs issued. Projects designated as workshops will include the development of RFPs to be issued subsequent to the workshops.

Watershed Management Workshop on *Cryptosporidium*

Review the current state of knowledge on sources of *Cryptosporidium* in watersheds, and develop relationships with nontraditional partners in planning watershed-level *Cryptosporidium* control measures.

Cryptosporidium Risk Checklist for Water Utilities

Synthesize results from related past and ongoing research, and develop a checklist identifying both watershed issues and treatment factors that may affect potential risks to a water utility from *Cryptosporidium*. Provide guidance to utilities for applying the checklist and interpreting the results once potential risks have been identified.

International Workshop on Support for New or Revised CT Tables for *Giardia, Cryptosporidium,* and Viruses

AWWARF will convene an international workshop in order to identify and evaluate data relevant to the latest research on *Giardia, Cryptosporidium,* and virus inactivation using alternate disinfectants such as ozone, chlorine dioxide, and chloramines. In addition, participants will identify and evaluate data relevant to inactivation of *Cryptosporidium* using chlorine.

Water Resources

Water Quality

Cyst and Oocyst Survival in Watersheds and Factors Affecting Inactivation [#151]

University of Ottawa, American Water Works Service Company, Regional Municipality of Ottawa-Carleton (Ont.), and City of Cornwall (Ont.)

Will evaluate the survival of *Cryptosporidium* oocysts and *Giardia* cysts exposed to differing environmental conditions and determine subsequent effects on disinfection efficiency. Will study effects of temperature, age, and physical stress on viability and susceptibility to disinfection. ***To be published in 1997.***

Effective Watershed Management for Surface Water Supplies [#317]

Portland (Ore.) Bureau of Water Works

Examines which watershed management practices best protect raw water supplies and documents installation, operation, and management costs of those judged technically and economically feasible for controlling THM precursors, general organics, iron, manganese, dissolved gases, algae, and algae nutrients. Also includes guidelines to help utility managers make watershed management decisions. Published in 1991. ***(Order 90587)***

Electronic Watershed Management Reference Manual [#903]

Camp Dresser & McKee, Inc.

Evaluates impacts of watershed protection measures on source water quality through quantitative evaluation and detailed analysis of best management practices. Develops guidance for utilities to maximize effectiveness in implementing these measures and identifies benefits within a regional and national regulatory framework. Also identifies federal, state, and select local regulatory programs involving watershed management practices. Published in 1995; *available only in CD-ROM format.* ***(Order 90695)***

Evaluation of Sources of Pathogens and NOM in Watersheds [#251]

Stroud Water Research Center, Wisconsin State Laboratory of Hygiene, South Central Connecticut Regional Water Authority (New Haven), and participating utilities

Will determine the distribution and densities of *Giardia* and *Cryptosporidium*, and concentrations of NOM in watersheds, and evaluate potential sources in field studies. Will develop potential source control strategies that will mitigate the concentrations of these contaminants in influent water resulting in potential treatment savings. ***To be published in 1998.***

Isotopic Tracers of Nonpoint-Source Pollution in Surface Waters [#376]

Lawrence Livermore National Laboratory

Evaluate existing isotope tracer techniques and develop additional techniques to specifically address nonpoint-source pollution in surface water in order to provide new data on nonpoint-source pollutants and the nature of their migration and fate in surface water. ***To be published in 1999.***

Water Treatment

Chemical Processes for Microbial, Organic, and Disinfection By-product (DBP) Control

Alternative Disinfection Technologies for Small Drinking Water Systems [#621]

Process Applications, Inc.

Summarizes the design requirements for the application of ozone, chlorine dioxide, and UV radiation for disinfection of small systems (pop. 1,000–10,000). Published in 1992. ***(Order 90619)***

Comprehensive Evaluation of *Cryptosporidium* Inactivation in Natural Waters [#375]

Southern Nevada Water Authority, Montgomery Watson, and participating utilities

Provide a mechanism and funding to publish the results of an ongoing nationwide study by Montgomery Watson that will determine inactivation of *Cryptosporidium* in natural waters using ozone and other disinfectants. ***To be published in 1999.***

Demonstration-Scale Evaluation of Engineering Aspects of the PEROXONE Advanced Oxidation Process [#525]

Metropolitan Water District of Southern California (Los Angeles) and Montgomery Watson

Will evaluate ozone and PEROXONE (ozone in combination with hydrogen peroxide) processes to confirm pilot-plant results for taste and odor control, disinfection by-product control, disinfection, and turbidity removal. Will also determine mass transfer efficiencies of the processes and evaluate various process considerations and equipment alternatives. ***To be published in 1997.***

Effect of Various Disinfection Methods on the Inactivation of *Cryptosporidium* [#906]

University of Alberta

Will investigate the effectiveness of ozone, potassium permanganate, chlorine, chlorine dioxide, and ultraviolet radiation (UV) on inactivation of *Cryptosporidium*. Also will investigate the use of combinations of disinfectants such as ozone–UV and ozone with hydrogen peroxide. ***To be published in 1997.***

Full-Scale Ozone Contactor Study [#630]

CH2M Hill Engineers, East Bay Municipal Utility District (Oakland, Calif.), Hackensack (N.J.) Water Company, and the Suburban Paris (France) Water Authority

Extensively evaluates five full-scale contactors (four types) to develop a model to predict contactor performance. Develops an approach for utilities to set operating rules for efficient operation while meeting *CT* requirements. Published in 1995. ***(Order 90668)***

Impact of Ozone on the Removal of Particles, TOC, and THM Precursors [#228]

University of North Carolina at Chapel Hill

Evaluates the performance of several full-scale plants using preozonation to limit THM formation. Also assesses the use of ozone as a coagulant and filter aid by determining ozone's effect on removal of particulates, TOC, and THM precursors by coagulation, sedimentation, and filtration. Published in 1989. ***(Order 90551)***

Improving Clearwell Design for *CT* Compliance [#271]

John Carollo Engineers, University of Illinois at Urbana-Champaign, Contra Costa Water District (Concord, Calif.), Alameda County (Calif.) Water District, City of Saint Louis (Mo.) Water Division, and Manitowoc (Wis.) Public Utilities

Will evaluate (1) various design configurations to retrofit or design clearwells with baffle systems and/or (2) modification of flow configurations that maximize retention time to meet *CT* disinfection requirements. ***To be published in 1998.***

Inactivation of *Giardia* Cysts With Chlorine at 0.5°C to 5°C [#107]

Colorado State University

Delineates amount of chlorine and contact time required to inactivate *Giardia* cysts at different disinfectant concentrations and pH ranges in low-temperature waters. These inactivation rates (*CT* values) were used to develop U.S. surface water treatment performance requirements. Published in 1987. ***(Out of print)***

Modeling Dissolved Ozone in Contactors [#632]

University of North Carolina at Chapel Hill and Los Angeles (Calif.) Dept. of Water and Power

Will develop a mathematical model describing ozone transfer and reactivity and dissolved ozone profiles in different types of contactors. Will verify the model at pilot and full scale and demonstrate its ability to evaluate disinfection efficiency using the *CT* concept. ***To be published in 1997.***

Ozone and Ozone–Peroxide Disinfection of *Giardia* and Viruses [#503]

University of Alberta

Evaluates disinfection performance criteria for disinfecting *Giardia* and viruses with ozone and ozone with hydrogen peroxide for waters of various qualities and temperatures. Presents a disinfection protocol for calculating *CT* (concentration × time) values that would account for advanced oxidation processes based on ozone. Published in 1992. ***(Order 90605)***

Ozone Disinfection of *Giardia* and *Cryptosporidium* [#731]

University of Alberta

Evaluates the relative resistance of *Giardia lamblia* cysts, *Giardia lamblia*, and *Cryptosporidium parvum* oocysts to inactivation by ozone in water at cold temperatures. Published in 1994. ***(Order 90661)***

Ozone in Water Treatment: Application and Engineering [#421]

AWWA Research Foundation and Compagnie Générale des Eaux

Describes current applications of ozone technology in drinking water treatment in terms of purpose (such as disinfection by-product control, taste and odor control, etc.), design, installation, and operation. Includes case studies and economic considerations. Published in 1991. ***Subscribers can order directly from the Research Foundation.*** Others can order from Lewis Publishers, Inc., by telephoning (800) 272-7737 for catalog no. L474LAFD.

Pilot-Scale Evaluation of Ozone and PEROXONE [#402]

Metropolitan Water District of Southern California (Los Angeles), University of California at Los Angeles, and James M. Montgomery Consulting Engineers

Investigates, by means of laboratory, pilot-plant, and field-scale testing, applications of the PEROXONE advanced oxidation process (ozone combined with hydrogen peroxide) for controlling disinfection by-products. Analyzes capital and operational costs of the process. Published in 1991. ***(Order 90591)***

Sequential Disinfection Design Criteria for the Inactivation of *Cryptosporidium* Oocysts in Drinking Water [#348]

University of Alberta

Building on work begun in project #273, will develop sufficient data on the inactivation of *Cryptosporidium* for estimation of the parameters for a kinetic model describing the synergistic effects of chemical disinfectants. Will also explore the effect of pH and temperature on the kinetic models. ***To be published in 1998.***

Synergistic Effects of Multiple Disinfectants [#273]

University of Alberta, City of Calgary (Alta.) Waterworks, Metropolitan Water District of Southern California (Los Angeles), and Regional Municipality of Waterloo (Ont.)

Will determine the presence of, and quantify the significance of, synergistic effects through the use of multiple disinfectant strategies. Will study the effects of sequential and simultaneous applications of chlorine, chlorine dioxide, monochloramine, UV, permanganate, and ozone on model organisms (*Giardia, Cryptosporidium,* and a bacterial system). ***To be published in 1998.***

Physical and Biological Processes for Microbial, Organic, and Disinfection By-product Control

Innovative Electrotechnologies for *Cryptosporidium* Inactivation [#282]

Clancy Environmental Consultants, Malcolm Pirnie, Inc., and University of Arizona

Will explore the potential of several innovative electrotechnologies for the inactivation of *Cryptosporidium* by means of a third party evaluation process. Will conduct three to four challenge demonstrations of innovative electrotechnologies. Will document power requirements and design considerations for the application of the technology in a range of sizes of water utility operations. Will include cyst seeding studies and inactivation determined by animal infectivity. This project is a collaborative effort of AWWARF and the Electric Power Research Institute Community Environmental Center. ***To be published in 1998.***

Integrated, Multi-Objective Membrane Systems for Control of Microbials and DBP Precursors [#264]

Kiwa N.V., University of Central Florida, Boyle Engineering Corporation, American Water Works Service Company, Irvine Ranch Water District (Calif.), City of Fort Myers (Fla.), Amsterdam (Netherlands) Water Supply, Water Supply of North Holland (Netherlands), and Water Supply Company (Overjessel, Netherlands)

Will optimize sequences of different membrane types—microfiltration (MF), ultrafiltration (UF), nanofiltration (NF), and reverse osmosis (RO)—that can function as a synergistic system for removing microbiological contaminants and DBP precursors. Will address the following important issues: use of staged membranes for pretreatment, minimization of chemical pretreatment, multiple treatment objectives, process sustainability, fouling minimization, reliability, and operational considerations. Will include development of a protocol for multiple membrane applications for surface water sources. ***To be published in 1999.***

Optimizing Filtration in Biological Filters [#252]

University of Waterloo, Metropolitan Water District of Southern California (Los Angeles), Georgia Institute of Technology, The Johns Hopkins University, and participating utilities

Will address key issues related to optimization of biological filters for multiple objectives, namely, simultaneous particle or floc removal and biodegradation of organic compounds, on a number of source waters and process designs (direct filtration, post-sedimentation, following ozone, etc.). ***To be published in 1998.***

Particulate and Microbial Removal

Characteristics of Initial Effluent Quality and its Implications for the Filter to Waste Procedure [#114]

Montana State University, City of Helena (Mont.), and City of Bozeman (Mont.)

Evaluates the validity of the filter to waste procedure by plant-scale investigation of initial filter effluent quality and pilot-plant study of filter-ripening mechanisms and backwashing techniques that improve initial effluent quality. Also describes quality of initial filtrate, presents guidelines for minimizing degradation of postbackwash filtered water quality, and discusses whether a filter to waste period for treatment plants is warranted. Published in 1988. ***(Out of print)***

Design and Operation Guidelines for Optimization of the High-Rate Filtration Process: Volume I—Plant Survey Results; Volume II—Plant Demonstration Studies [#303]

Iowa State University

Develops guidelines for designing and operating high-rate filtration plants capable of producing high-quality finished water (ntu less than 0.2), based on plant-scale validation of key parameters identified through a survey of existing high-rate plants and pilot-plant research. Also provides guidelines for modifying processes and operations of existing plants to achieve high-quality finished water and estimate attendant costs. Results of survey phase published in 1989. ***(Order 90552)*** Results of pilot and full scale phase published in 1992. ***(Order 90596)***

Dissolved Air Flotation: Laboratory and Pilot Plant Investigations [#536]

University of Massachusetts and South Central Connecticut Regional Water Authority (New Haven)

Investigates dissolved air flotation (DAF) as an alternative to gravity settling. In the first phase, relates air requirements to source water quality and pretreatment conditions, and compares DAF performance with conventional treatment on a lab and pilot scale. In the second phase, evaluates performance of DAF facilities in North America and Europe, and develops cost data for this process relative to conventional treatment. Results of first phase published in 1992 as *Dissolved Air Flotation: Laboratory and Pilot Plant Investigations.* ***(Order 90611)*** Results of second phase published in 1994 as *Dissolved Air Flotation: Field Investigations.* ***(Order 90651)***

Enhanced and Optimized Coagulation for Removal of Particulates and Microbial Contaminants [#155]

American Water Works Service Company, University of Colorado, University of Washington, The Johns Hopkins University, Manatee County (Fla.) Utilities Dept., City of Grand Forks (N.D.), City of Martinez (Calif.), Dallas (Texas) Water Utilities, and Houston (Texas) Dept. of Public Works

Will evaluate the effect of enhanced and optimized coagulation, as defined by the pending D/DBP rule, on particulate and microbial removal. Will include bench-scale studies on 18 waters corresponding to the USEPA TOC–alkalinity matrix and removal of protozoan cysts and oocysts, viruses, enteric bacteria, spores, and bacteriophage. Will perform larger scale testing using similar parameters and develop operational guidelines for utility application. ***To be published in 1997.***

Evaluation of Ultrafiltration Membrane Pretreatment and Nanofiltration of Surface Waters [#601]

James M. Montgomery Consulting Engineers, East Bay Municipal Utility District (Oakland, Calif.), Contra Costa Water District (Concord, Calif.), Lyonnaise des Eaux-Dumez (France)

Evaluates the impact of pre- and post-treatment on low-pressure membrane processes to determine the effect on operation, removal of DBP precursors, tastes and odors, and corrosion control. Develops cost comparisons and analysis for different membrane process schemes. Published in 1994. ***(Order 90639)***

Filtration of *Giardia* Cysts and Other Particles Under Treatment Plant Conditions [#80]

Colorado State University, City of Fort Collins (Colo.), and Town of Empire (Colo.)

Delineates conditions that have contributed to or permitted passage of *Giardia* cysts by the rapid-rate filtration process. Evaluates applicability to full-scale operation of the findings of recent pilot-scale studies of rapid-rate filtration (with special reference to low-turbidity water). Also evaluates startup and operation of a newly constructed slow sand filtration system in a watershed known to contain a persistent source of *Giardia* cysts. Published in 1988. ***(Order 90533)***

Full-Scale Evaluation of Declining and Constant Rate Filtration [#202]

Environmental Engineering & Technology, Inc., and City of Durham (N.C.)

Compares effluent quality relative to benefits of constant- and declining-rate filtration control methods by evaluating them under a variety of operating conditions at a full-scale plant with split-flow capability. Published in 1992. ***(Order 90579)***

Giardia and *Cryptosporidium* in Water Supplies [#430]

American Water Works Service Company and University of Arizona

Evaluates more than 60 surface water treatment filter plants in North America for the occurrence and distribution of *Giardia* and *Cryptosporidium* in raw and treated surface waters to determine the occurrence of these parasites and the efficacy of various treatment practices in their removal. Published in 1991. ***(Order 90583)***

Low Pressure Membrane Filtration for Particle Removal [#506]

Boise (Idaho) Water Corporation

Demonstrates the applicability of ultrafiltration membrane separation to water treatment and provides an integrated technical and economic evaluation of the process. Evaluates the cost, efficiency, limitations, and operation and maintenance requirements of the process under a variety of real-world conditions. Published in 1992. ***(Order 90603)***

Manual of Design for Slow Sand Filtration [#404]

Colorado State University, CH2M Hill Engineers, Denver (Colo.) Water Dept., the State of Colorado, Black & Veatch, Dayton & Knight, and RBD Consulting Engineers

Describes engineering, design, and construction practices necessary to achieve desired capacity and reliability of slow sand systems given anticipated source water quality conditions. Emphasizes using locally available building components, construction materials, and contractors. Published in 1991. ***(Order 90578)***

Membrane Filtration for Microbial Removal [#817]

Montgomery Watson

Will determine the efficacy of available membrane technologies for removing microorganisms to meet regulatory compliance requirements of the Surface Water Treatment Rule and the forthcoming Ground Water Disinfection Rule. Will elucidate the mechanism for microbial removal and develop an economic cost model. Published in 1996. ***(Order 90715)***

Mixing in Coagulation and Flocculation [#316]

AWWA Research Foundation international authors committee

Reports on the theory, design, and practice of mixing in coagulation and flocculation by experts from both the chemical engineering and water treatment fields. Published in 1991. ***(Order 90580)***

National Assessment of Particle Removals by Filtration [#908]

Environmental Engineering & Technology, Inc., American Water Works Service Company, and Erie County (N.Y.) Water Authority

Will determine the range of particles in filtered effluent as a benchmark for water utilities to use in evaluating treatment performance. Will generate the particle count data via an extensive sampling program in which samples will be collected on a seasonal basis from a number of U.S. treatment plants. Will provide utilities with particle count information to which they can compare their individual performance. ***To be published in 1997.***

Optimization of Filtration for Cyst Removal [#703]

Montgomery Watson, Portland (Ore.) Bureau of Water Works, Contra Costa Water District (Concord, Calif.), City of Seattle (Wash.), Tualatin Valley Water District (Wash.), Azusa Valley Water Company (Calif.), City of Pomona (Calif.), City of Pasadena (Calif.), City of Ontario (Calif.), and Covina Irrigation

Will investigate *Cryptosporidium* and *Giardia* removal efficiencies in a variety of pilot-scale filtration systems. Will study treatment variables such as coagulant type and dose, physical mixing processes, pH adjustment (ahead of filtration), filter aids, and filter media. Will also evaluate operational variables such as filter start-up, filter loading rates, length of run, backwash type, and backwash frequency. Published in 1996. ***(Order 90699)***

Optimizing Treatment for Pathogen Removal [#274]

Will report on an AWWARF expert workshop. The objective of the workshop was to provide information on administrative, operational, and technical factors affecting plant optimization for particle (*Cryptosporidium*) removal at conventional treatment plants. The information developed in the workshop has since been represented in the Partnership for Safe Water Self Assessment. The self-assessment tool was designed to be used by utilities to evaluate and improve treatment, primarily for removal of waterborne pathogens. In addition to this use of the AWWARF workshop results, the Foundation will develop a document that will identify factors and provide guidance related to treatment plant optimization. ***To be published early in 1997.***

Procedures Manual for Polymer Selection in Water Treatment Plants [#209]

University of Delaware

Based on utility use of the protocol published for project #81, the manual has been revised. Includes two new modules on polymer charge density and molecular weight (intrinsic viscosity) determination, deletes the module on bench-scale filters, and improves the remaining modules. Published in 1989. ***(Order 90553)***

The Removal and Disinfection Efficiency of Lime Softening Processes for *Giardia* and Viruses [#608]

Black & Veatch

Determines the extent of inactivation/removal credit for *Giardia* cysts and viruses that can be given to utilities that use lime softening to treat surface water or groundwater under the direct influence of surface water. Includes bench-scale studies on softened waters at various pH conditions, with and without chemical disinfectant present. Published in 1994. ***(Order 90648)***

Selection and Design of Mixing Processes for Coagulation [#613]

University of Illinois at Urbana–Champaign, Contra Costa Water District (Concord, Calif.), Los Angeles (Calif.) Dept. of Water and Power, and James M. Montgomery Consulting Engineers

Examines the effect of rapid (flash) mixing on the efficiency of downstream processes, i.e., flocculation, sedimentation, and gravity filtration. Also determines fundamental mixing characteristics of different rapid mix technologies and addresses scale-up issues. Provides practical design approaches for both rapid mixing and flocculation. Published in 1994. ***(Order 90641)***

A Study of Water Treatment Practices for the Removal of *Giardia lamblia* Cysts [#74]

University of Washington

Evaluates the ability of full-scale filtration systems (rapid sand, dual media, and diatomaceous earth) operating under ambient conditions to remove *Giardia* cysts and cyst-size particulates. Provides design and operation guidelines for maximizing cyst removal and verifies findings of previous laboratory and pilot-plant studies on cyst-removal techniques. Published in 1989. ***(Out of print)***

Water Treatment Residuals

Recycle Stream Effects on Water Treatment [#514]

Environmental Engineering & Technology, Inc., and American Water Works Service Company

Studies the beneficial and adverse effects of sidestream recycling. Evaluates the impact of recycling practices on coagulation, filtration, and disinfection by-products. Covers a broad range of recycling options in laboratory- and full-scale research. Published in 1993. ***(Order 90629)***

Treatment Options for *Giardia, Cryptosporidium,* and Other Contaminants in Recycled Backwash Water [#352]

Environmental Engineering & Technology, Inc. and American Water Works Service Company

Evaluate various treatment and process control options with the goal of identifying cost-effective methods for reducing cysts and oocysts in recycled backwash water. Integrate cyst and oocyst removal with treatment for other contaminants that may impair finished water quality. ***To be published in 1999.***

Distribution Systems

Maintenance of Water Quality

Booster Disinfection to Improve Water Quality [#261]

University of Cincinnati, USEPA, Irvine Ranch Water District (Calif.), Cincinnati (Ohio) Dept. of Water, City of Midland (Texas), Portland (Ore.) Bureau of Water Works, City of New York (N.Y.) Dept. of Environmental Protection, and Second Taxing District (Conn.) Water Dept.

Will investigate techniques and develop a manual of best practice for the location and operation of booster disinfection systems. Will aid utilities in compliance with the Total Coliform Rule, D/DBP Rule, and Ground Water Disinfection Rule. ***To be published in 1998.***

Monitoring and Analysis

Methods Development

Application of Surrogate Measures to Improve Plant Performance [#363]

Utah State University, CH2M Hill Engineers, and participating utilities

Evaluate the use of pathogen surrogates (e.g., particle counts, aerobic spores, algal cells, *Clostridium* spores) for improving plant performance. Develop correlations between surrogates or pathogens and common online measures of performance (e.g., turbidity, particle counts) under full-scale conditions, and quantify these correlations through the individual plant processes for a range of source waters and treatment processes. ***To be published in 1999.***

Biological Particle Surrogates for Filtration Performance Evaluation [#181]

Colorado State University and City of Bellingham (Wash.)

Will determine species of biological particles that mimic the behavior of pathogenic species such as viruses, bacteria, and protozoan cysts in coagulation and filtration. Will also ascertain relationships between removals of test particles and turbidity and particle removals. ***To be published in 1997.***

Cryptosporidium and *Giardia* Antibody Protocol [#358]

Wisconsin State Laboratory of Hygiene, Indiana University–Kokomo, City of Milwaukee (Wis.), City of Oshkosh (Wis.), Windsor Utilities Commission (Ont.), and City of Appleton (Wis.) Water Utility

Develop a protocol, including a mechanism to determine equivalency and improvements needed, to evaluate various antibodies that could be used for detection of *Giardia* and *Cryptosporidium*. ***To be published in 1999.***

Cryptosporidium parvum Viability Assay [#351]

University of Alberta

Develop a method to determine viability of *Cryptosporidium parvum* in finished drinking water. ***To be published in 1999.***

Cryptosporidium Viability Study [#395]

Clancy Envrionmental Consultants, Inc., Thames Water Utilities Ltd. (U.K.), University of Arizona, and Scottish Parasite Diagnostic Laboratory (U.K.)

Will compare at least three different methods for determining the viability and infectivity of oocyts: DAPI–PI vital stain, mouse infectivity, and other vital dyes. Will use strict quality assurance conditions to ensure meaningful comparisons and will consider oocyst age and effects of treatment stresses. The United Kingdom Drinking Water Inspectorate is co-funding this project. ***To be published in 1999.***

Detection of *Cryptosporidium* and *Giardia* by Flow Cytometry [#283]

Wisconsin State Laboratory of Hygiene

Will evaluate the use of flow cytometry to improve detection of *Giardia* and *Cryptosporidium* using the immunofluorescence assay. ***To be published in 1997.***

Development of a Test to Assess *Cryptosporidium parvum* Oocysts Viability: Correlation With Infectivity Potential [#609]

University of Arizona

Develops and field tests a method for determining *C. parvum* viability that can be related to infectivity potential. Includes use of vital dyes, differential interference contrast microscopy, and phase contrast microscopy. Published in 1993. ***(Order 90626)***

Improving the Immunofluorescence Assay (IFA) Method for *Giardia* and *Cryptosporidium* [#259]

New York State Dept. of Health—Wadsworth Laboratories, Health Research, Inc., Erie County (N.Y.) Water Authority, City of Troy (N.Y.), Latham (N.Y.) Water District, City of Poughkeepsie (N.Y.)

Will address the limitations of the current ICR method, including filtration, concentration, flotation, and identification to increase recoveries and reproducibility. Will compare modifications to this method with the current method and include limited field testing. ***To be published in 1998.***

Incorporation of a Vital Stain for *Giardia* and *Cryptosporidium* Into the Immunofluorescence Assay [#160]

University of Alberta and City of Edmonton (Alta.)

Will search for a vital stain for *Giardia* cysts and *Cryptosporidium* oocysts that can be incorporated into the immunofluorescence assay (IFA) for drinking water. Will investigate sources of error and interferences and their impact on the IFA. Also will test the vital stain in several natural waters and its "ruggedness" in several different laboratories. ***To be published in 1997.***

Method for *Giardia* and *Cryptosporidium* Detection and Viability [#253]

Thames Water Utilities Ltd. (U.K.), Scottish Parasite Diagnostic Laboratory, Huntington Water Treatment (U.K.), Erie County (N.Y.) Water Authority, and Saint Louis County (Mo.) Water Company

Will evaluate methods that improve *Giardia* cyst and *Cryptosporidium* oocyst recovery and identification efficiencies and that enable simultaneous viability determination. Will also identify simpler, rapid methods that require less technical expertise than those methods currently available and that are less labor intensive. ***To be published in 1998.***

New Approaches for Isolation of *Cryptosporidium* and *Giardia* [#364]

Clancy Environmental Consultants, Inc. and Thames Water Utilities Ltd. (U.K.)

Develop new application to optimize the recovery of *Giardia* cysts and *Cryptosporidium* oocysts from raw and finished waters. Evaluate alternative sample collection and concentration techniques and novel purification and separation methods. ***To be published in 2000.***

UV–VIS Spectroscopy for the Rapid On-Line Detection of Protozoa [#162]

University of South Florida, Green Bay (Wis.) Water Utility, and Tampa (Fla.) Water Dept.

Will develop instrumentation based on ultraviolet-visible (UV–VIS) spectroscopy for use as an on-line particle analyzer to detect, identify, and quantify *Cryptosporidium* oocysts and *Giardia* cysts. ***To be published in 1998.***

Automated Monitoring and Instrumentation

Evaluation of Particle Counting as a Measure of Treatment Plant Performance [#505]

City of Calgary (Alta.) Waterworks and Bigelow Laboratory for Ocean Sciences

Assesses particle size analysis as a means of optimizing the filtration process. Evaluates sampling protocols and commercially available particle-counting instruments. Includes a pilot-scale study of pathogen removal and an evaluation of filtration processes. Published in 1992. ***(Order 90595)***

A Practical Guide to On-Line Particle Counting [#835]

City of Calgary (Alta.) Waterworks

Provides information for utilities to put particle counters on-line. Focuses on six areas: sampling configuration, sensor flow control, particle concentration and size issues, sensor intercalibration, data presentation, and quality assurance. Published in 1995. ***(Order 90674)***

Quantitative Particle Count Method Development: Count Standardization and Sample Stability/Shipping Considerations [#266]

Malcolm Pirnie, Inc., University of Texas, City of Chandler (Ariz.), Los Angeles (Calif.) Dept. of Water and Power, City of Everett (Wash.), City of San Diego (Calif.), City of Newport News (Va.), and Waco (Texas) Water Treatment Plant

Will evaluate sample stability and the potential for standardized quantitative particle count analysis for drinking water quality monitoring. Will provide broader access of particle counting technology to water utilities. ***To be published in 1998.***

Management and Administration

Organization Structure

Managerial Assessment for Water Quality and System Reliability [#257]

Apogee Research, Inc., Malcolm Pirnie, Inc., Metropolitan Water District of Southern California (Los Angeles), and participating utilities

Will use an American Assembly approach to convene water utility managers, public works directors, and plant superintendents to produce a consensus document for the managerial assessment of treatment reliability. Will provide a holistic source-to-tap framework for organizing the measures to assure reliability that can serve as benchmarks to anticipate future needs and strategic initiatives. Will include a 14½ min. VHS video. ***To be published in 1997.***

Public Education and Communication

Synthesis Report on *Cryptosporidium* [#372]

Black & Veatch

Summarize the results of international *Cryptosporidium* research, highlighting AWWARF studies, into one document. ***To be published in 1997.***

Health Effects

Risk Assessment Reviews

Microbial Risk Assessment [#281]

International Life Sciences Institute

Will report on a microbial risk assessment model of waterborne contaminants for food and water evaluations. Will account for qualitative and quantitative information associated with exposure to either a microbiological organism or the media in which the organism occurs. Will identify research needs for analytical methods. The AWWA Research Foundation is a co-sponsor of this project. ***To be published in 1997.***

Microbial Risk Assessment for Drinking Water [#801]

University of South Florida

Will characterize the risks of pathogenic microorganisms that may be transmitted through drinking water. Will also develop quantitative risk assessment models based upon human dose–response data for a key group of waterborne pathogens. Will address factors such as infectious dose, ratio of clinical to subclinical infections, variations in microbial virulence, host immunity, and mortality rates. ***To be published in 1997.***

Workshop to Examine the Role of Drinking Water in Cryptosporidiosis in the Immunocompromised Population: Review of Completed Studies and Design of Future Study [#377]

Review ongoing and completed health risk studies to define the current understanding of the role of drinking water in the transmission of cryptosporidiosis in the AIDS population and design an epidemiologic research program to quantify the risk. ***Internal report available upon request.***

Laboratory and Field Studies

Assessment of Molecular Epidemiology of Waterborne *Cryptosporidium* With Respect to Origin [#366]

Metropolitan Water District of Southern California (Los Angeles), University of California–Davis, City of Los Angeles (Calif.) Department of Water and Power, Alameda County (Calif.) Water District, Orange County (Calif.) Water District, and Modesto (Calif.) Irrigation District

Determine the strain variability of *Cryptosporidium parvum* oocysts with respect to origin using a molecular fingerprinting approach developed from available polymerase chain reaction technology. ***To be published in 1999.***

Association Between Occurrence of *Cryptosporidium* in Finished Water and Cryptosporidiosis in the Population Served [#177]

The Lovelace Institutes

Will examine the occurrence of cryptosporidiosis in a population consuming groundwater with that of one consuming surface water. Using a comparison of the relative infection rates, will provide an estimate of the relative contribution of water to infection rates. ***To be published in 1997.***

Cryptosporidium parvum: Surrogate Human Pathogenicity Animal Model Using Swine—Phase I [#344]

Utah State University

Establish and validate an animal model in pigs for application to human risks for infectivity and virulence of *Cryptosporidium* and possible other pathogenic organism field isolates, including the quantitative examination of variables such as immunocompetency and age. ***To be published in 1999.***

Technology Transfer

Videotapes

The AWWA Research Foundation produces videotapes to communicate the results of its research projects.

Currently, fifteen videotapes are available on loan. They include:

(1) *Controlling* Giardia lamblia describes five projects on *Giardia.*

(2) *The Ozone Option* discusses various ozone projects.

(3) *The Balancing Act: Disinfection and Disinfection By-product Control* provides an overview of the issues involved in balancing the trade-offs between disinfection and disinfection by-product control.

(4) *Disinfection: Weighing the Alternatives,* also a companion to *The Balancing Act* video, focuses on disinfectants and treatment techniques for the inactivation of microorganisms.

These productions, ranging in length from 11 to 20 minutes, are available in ½-inch VHS format. They may be borrowed at no charge by calling (303) 347-6211.

REFERENCES

Anderson, W.L., T.L. Champlin, W.F. Clunie, D.W. Hendricks, D.A. Klein, P. Kugrens, and G. Sturbaum. 1996. Biological Particle Surrogates for Filtration Performance Evaluation. Toronto, Ontario: *AWWA ACE Proceedings*.

APHA, AWWA, and WEF (American Public Health Association, American Water Works Association, and Water Environment Federation). 1994. Section 9711 B. Immunofluorescence Method for *Giardia* and *Cryptosporidium* spp. (Proposed) & Section 2560 Particle Counting and Size Distribution (Proposed). *Standard Methods for the Examination of Water and Wastewater.* 18th ed. Washington, D.C.: APHA, pp. 64-70 and pp.1-9.

AWWA (American Water Works Association). 1995. *What Water Utilities Can Do to Minimize Public Exposure to* Cryptosporidium *in Drinking Water*. White Paper. Denver, Colo.: AWWA.

AWWARF (American Water Works Association, Research Foundation). 1997. AWWARF American Assembly Statement. In J. Cromwell, D. Owen, J. Dyksen, and E. Means. *Managerial Assessment of Water Quality and System Reliability.* Denver, Colo.: AWWA and AWWARF.

Badenoch, J. 1995. *Cryptosporidium* in Water Supplies. Report of the Group of Experts. London: Department of the Environment, Department of Health, Her Majesty's Stationery Office.

Carrington, E.G., and M.E. Ransome. 1994. Factors Influencing the Survival of *Cryptosporidium* Oocysts in the Environment. [Report:FR0456] Marlow, England: Foundation for Water Research.

CDC (Centers for Disease Control). 1986. Epidemiologic Notes and Reports, Cryptosporidiosis — New Mexico. *Morbidity and Mortality Weekly Rept.*, 36(33):561.

———. 1993. Surveillance for Waterborne Disease Outbreaks — U.S. — 1991-1992. *Morbidity and Mortality Weekly Rept.,* 42(55-5):1.

———. 1995. Assessing the Public Health Threat Associated With Waterborne Cryptosporidiosis: Report of a Workshop. *Morbidity and Mortality Weekly,* Vol 44 (RR-6) : June.

Clancy, J.L., W.D. Gollintz, and Z. Tabib. 1994. Commercial Labs: How Accurate Are They? *Jour. AWWA,* 86(5):89-97.

Clancy, J. L., R.M. McCuin, and S. Schaub. 1995. *USEPA Performance Evaluation of the ICR Method for* Giardia *and* Cryptosporidium: *A Report*. Washington, D.C.: USEPA.

Clancy, J.L. 1996a. *Performance Evaluation of Canadian Laboratories — Recovery and Enumeration of* Giardia *Cysts and* Cryptosporidium *Oocysts From Water Samples.* Project Number k221366. Weston, Ontario: Health Canada.

Clancy, J.L. 1996b. *Evaluation of the Proposed ICR Protocol for Determining the Field Recovery of Protozoan Cysts From Source and Treated Drinking Water.* 68-C5-3909:26. Washington, D.C.: USEPA, Office of Science and Technology.

Cryptosporidium Capsule. 1996. Largest Recreational Waterborne Outbreak — Collingwood Boil Water Advisory Still in Place. 1(8):10-11.

D'Antonio, R.G., R.E. Winn, J.P. Taylor, T.L. Gustafson, W.L. Current, M.M. Rhodes, G.W. Gary, and R.A. Zajac. 1985. Waterborne Outbreak of Cryptosporidiosis in Normal Hosts. *Ann. Internal Med.,* 103(b):886.

Fayer, R., and B.L.P. Ungar. 1986. *Cryptosporidium* spp. and Cryptosporidiosis. *Microbiological Reviews,* 50(4):458-483.

Fayer, R., C. Andrews, B.L.P. Ungar, and B. Blagburn. 1989. Efficacy of Hyperimmune Bovine Colostrum for Prophylaxis of Cryptosporidiosis in Neonatal Calves. *Jour. Parasitology,* 75:393.

Fayer, R. 1996. Introduction to the *Cryptosporidium* Issue. In *Proc. of AWWA ACE Toronto, Ontario.* Denver, Colo.: AWWA.

Fayer, R., T.K. Graczyk, C.A. Farley, E.J. Lewis, and J.M. Trout. 1997. Potential Role of Water Fowl and Oysters in Complex Epidemiology of *Cryptosporidium Parvum.* In *Proc. of* the International Symposium on *Cryptosporidium* in Drinking Water, Newport Beach, CA, March 3–7.

Finch, G.R., E.K. Black, L. Gyürek, and M. Belosevic. 1994. *Ozone Disinfection of* Giardia *and* Cryptosporidium. Denver, Colo.: AWWARF and AWWA.

Fox, K.R., and D.A. Lytle. 1996. Milwaukee's Cryptosporidiosis Outbreak: Investigation and Recommendations. *Jour. AWWA,* 883(9):87-94.

Frey, M.M., C.D. Hancock, and G.S. Logsdon. 1997. *Critical Evaluation of* Cryptosporidium *Research and Research Needs*. Denver, Colo.: AWWARF and AWWA.

Frost, F.J., G.F. Craun, and R.L. Calderon. 1996. Waterborne Disease Surveillance. *Jour. AWWA,* 88(9):66-75.

Goldstein, S. 1995. *An Outbreak of Cryptosporidiosis in Clark County, Nevada: Summary of Investigation.* Epi-aid #94-45-1. Atlanta, Ga.: Centers for Disease Control.

Hansen, J.S., and J.E. Ongerth. 1991. Effects of Time and Watershed Characteristics on the Concentration of *Cryptosporidium* Oocysts in River Water. *App. Env. Microbio.,* 57(10):2790-2795.

Hayes, E.B., T.D. Matte, T.R. O'Brien, T.W. McKinley, G.S Logsdon, J.B. Rose, B.L.P. Ungar, D.M Word, P.F. Pinsky, M.L. Cummings, M.A. Wilson, E.G. Long, E.S. Hurwitz, and D.D. Juranek. 1989. Large Community Outbreak of Cryptosporidiosis Due to Contamination of a Filtered Public Water Supply. *N.E. Jour. of Medicine,* 320(21):1372-1376.

Jacangelo, J.G., S.S. Adham, and J.M. Laîné. 1995. Mechanisms of *Cryptosporidium*, *Giardia* and MS2 Virus Removal by MF and UF. *Jour. AWWA,* 87(9):107-121.

Kelley, M.B., J.K. Brokaw, J.K. Edzwald, D.W. Fredericksen, and P.K. Warrier. 1994. A Survey of Eastern U.S. Army Installation Drinking Water Sources and Treatment Systems for *Giardia* and *Cryptosporidium*. Paper presented at AWWA Water Quality and Technology Conference, San Francisco, Calif.

Korich, D.G., J.R. Mead, J.S. Madore, N.A. Sinclair, and C.R. Sterling. 1990. Effects of Ozone, Chlorine Dioxide, Chlorine and Monochloramine on *Cryptosporidium* Oocyst Viability. *App. Env. Microbio.,* 56:1423-1428.

LeChevallier, M.W., W.D. Norton, and R.G. Lee. 1991a. Occurrence of *Giardia* and *Cryptosporidium* spp. in Surface Water Supplies. *App. Env. Microbio.,* 57(9):2610-2616.

LeChevallier, M.W., W.D. Norton, R.G. Lee, and J.B. Rose. 1991b. Giardia *and* Cryptosporidium *in Water Supplies*. Denver, Colo.: AWWARF and AWWA.

Lisle, J.T., and J.B. Rose. 1995. *Cryptosporidium* Contamination of Water in the U.S.A. and U.K.: A Mini-Review. *Jour. Water SRT-Aqua,* 44(3):103-117.

Madore, M.S., J.B. Rose, C.P. Gerba, M.J. Arrowood, and C.R. Sterling. 1987. Occurrence of *Cryptosporidium* Oocysts in Sewage Effluents and Select Surface Waters. *Jour. Parasitology,* 73:702-705.

Minnesota Department of Health. 1996. An Outbreak of Cryptosporidiosis at a Lake Resort, Cook County, Minnesota (draft). Minneapolis, Minn.

National *Cryptosporidium* Survey Group. 1992. A Survey of *Cryptosporidium* Oocysts in Surface and Groundwaters in the U.K. *Jour. Int. Water Engng. Mgmt.,* 6:697.

Nieminski, E.C. 1995. *Giardia* and *Cryptosporidium* Cysts Removal Through Direct Filtration and Conventional Treatment. Paper presented at Annual AWWA Conf., New York, June 19-23.

Nieminski, E.C., and J.E. Ongerth. 1995. Removing *Giardia* and *Cryptosporidium* by Conventional Treatment and Direct Filtration. *Jour. AWWA,* 87(9):96-106.

Ong, C., W. Moorehead, A. Ross, and J. Isaac-Renton. 1996. Studies of *Giardia* spp. and *Cryptosporidium* spp. in Two Adjacent Watersheds. *App. Env. Microbio.,* 62(8):2798-2805.

Ongerth, J.E., and H.H. Stibbs. 1987. Identification of *Cryptosporidium* Oocysts in River Water. *App. Env. Microbio.,* 53(4):672-676.

Oregon Health Division. 1992. A Large Outbreak of Cryptosporidiosis in Jackson County. *Communicable Disease Summ.,* 41(July 14):14.

Owens, J.H., R.J. Miltner, F.W. Schaefer, and E.W. Rice. 1994. Pilot-Scale Ozone Inactivation of *Cryptosporidium* and *Giardia*. In *Proc. of AWWA Water Quality and Technology Conference, San Francisco, Calif.*

Parker, J.F.W., G.F. Greaves, and H.V. Smith. 1992. The Effect of Ozone on the Viability of *Cryptosporidium* Parvum Oocysts and a Comparison of Experimental Methods. *Water Sci. Technol.,* 27:93-96.

Patania, N.L., J.G. Jacangelo, L. Cummings, A. Wilczak, K. Riley, and J. Oppenheimer. 1995. *Optimization of Filtration for Cyst Removal.* Denver, Colo.: AWWARF and AWWA.

Plummer, J.D., J.K. Edzwald, and M.B. Kelley. 1995. Removing *Cryptosporidium* by Dissolved-Air Flotation. *Jour. AWWA,* 87(9):85-95.

Richardson, A.J., R.A. Frankenberg, A.C. Buck, J.B. Selkon, J.S. Coulbourne, J.W. Parsons, and R.T. Mayon-White. 1991. An Outbreak of Waterborne *Cryptosporidiosis* in Swindon and Oxfordshire. *Epidemiol. Infect.*, 107:485-495.

Roach, P.D., M.E. Olson, G. Whitley, and P.M. Wallis. 1993. Waterborne *Giardia* Cysts and *Cryptosporidium* Oocysts in the Yukon, Canada. *App. Envir. Microbio.,* 59(1):67-73.

Robertson, L.J., A.T. Campbell, and H.V. Smith. 1992. Survival of Oocysts of *Cryptosporidium parvum* Under Various Environmental Pressures. *App. Env. Microbio.,* 58:3494-3500.

Rose, J.B., H. Darbin, and C.P. Gerba. 1988a. Occurrence and Significance of *Cryptosporidium* in Water. *Jour. AWWA,* 80(2):53-58.

Rose, J.B., H. Darbin, and C.P. Gerba. 1988b. Correlations of the Protozoa *Cryptosporidium* and *Giardia* With Water Quality Variables in a Watershed. *Water Science and Technology,* 20:271-276.

Rose, J.B., C.P. Gerba, and W. Jakubowski. 1991. Survey of Potable Water Supplies for *Cryptosporidium* and *Giardia*. *Environ. Sci. Tech.,* 25:1393-1400.

Rush, B.A., P.A. Chapman, and R.W. Inesan. 1990. A Probable Waterborne Outbreak of Cryptosporidiosis in the Sheffield Area. *Jour. Med. Microbiol.,* 321:239-242.

Schuler, P.F., M.M. Ghosh, and S.N. Boutros. 1988. Comparing the Removal of *Giardia* and *Cryptosporidium* Using Slow Sand and Diatomaceous Earth Filtration. In *Proc. of AWWA ACE.* Denver, Colo.: AWWA.

Smith, H.V., W.J. Patterson, R. Hardie, L.A. Greene, C. Bentos, W. Tulloch, R. A. Gilmour, R.W.A. Girdwood, J.C.M. Sharp, and G.L. Forbes. 1989. An Outbreak of Waterborne Cryptosporidiosis Caused by Post-Treatment Contamination. *Epidem. Infect.*, 103:703-715.

Solo-Gabriele, H., and S. Neumeister. 1996. U.S. Outbreaks of Cryptosporidiosis. *Jour. AWWA,* 88(9):76-85.

U.S. EPA (United States Environmental Protection Agency). 1996. National Primary Drinking Water Regulations: Monitoring Requirements for Public Drinking Water Supplies; Final Rule. *Federal Register*, 61(94):24354-24388.

Working Group on Waterborne Cryptosporidiosis. 1997. Cryptosporidium *and Water: A Public Health Handbook.* Atlanta, Ga.: Council of State and Territorial Epidemiologists (also distributed through AWWA, Denver, Colo.).

ABBREVIATIONS

AIDS	acquired immunodeficiency syndrome
APHA	American Public Health Association
AWWA	American Water Works Association
AWWARF	American Water Works Association Research Foundation
°C	degrees Celsius
CDC	Centers for Disease Control
CT	disinfectant residual concentration multiplied by contact time
CV	coefficient of variation
DAF	dissolved air flotation
DE	diatomaceous earth
DEP	Department of Environmental Protection (New York City)
EPA	Environmental Protection Agency (U.S.)
ICR	Information Collection Rule
IESWTR	Interim Enhanced Surface Water Treatment Rule
IFA	immunofluorescence assay
L	liter
log	logarithm
M$	million dollars
mg	milligram
min	minute
mL	milliliter
μm	micrometer
mm	millimeter
n	number of samples
NA	information not available
pH	negative logarithm of the effective hydrogen ion concentration
SDWA	Safe Drinking Water Act

SD	standard deviation
SWTR	Surface Water Treatment Rule
UK	United Kingdom
U.S.	United States
U.S. EPA	United States Environmental Protection Agency
UV	ultraviolet
WEF	Water Environment Federation
WTP	water treatment plant
WWW	World Wide Web